SpringerBriefs in Geography

For further volumes:
http://www.springer.com/series/10050

Basudeb Bhatta

Urban Growth Analysis and Remote Sensing

A Case Study of Kolkata, India 1980–2010

Springer

Basudeb Bhatta
Computer Aided Design Centre
Department of Computer Science and Engineering
Jadavpur University
Kolkata
700032
India

ISSN 2211-4165 ISSN 2211-4173 (electronic)
ISBN 978-94-007-4697-8 ISBN 978-94-007-4698-5 (eBook)
DOI 10.1007/978-94-007-4698-5
Springer Dordrecht Heidelberg New York London

Library of Congress Control Number: 2012939655

Printed on acid-free paper

Springer is part of Springer Science+Business Media (www.springer.com)

Dedicated to all my teachers

Preface

Study of urban growth is a branch of urban geography that concentrates on cities and towns in terms of their physical and demographic expansion. Understanding the urban patterns, dynamic processes, and their relationships is a primary objective in the urban research agenda with a wide consensus among scientists, resource managers, and planners; because future development and management of urban areas requires detailed information about ongoing processes and patterns. Analysis of urban patterns and processes, from remote sensing data, is a pertinent topic in the current urban research agenda. Detailed spatial and temporal information of urban morphology, infrastructure, population distributions, land-use/land-cover patterns and transitions among different land-uses/land covers are essential to be observed and understood. Urban remote sensing has attempted to provide such information. Determining the rate and the spatial configuration of urban growth from remote sensing data is not only a prevalent approach, rather it has a long history. However, the models and methods applied on remote sensing data, in terms of urban growth analysis, differ widely. Scientists and researchers hold different opinions. Many of them are still working to develop new methods and robust tools. This book is to document one such research.

This book aims to demonstrate some of the existing methods/models to test their fitness for the analysis of urban growth in the city of Kolkata, an important urban area from a developing country. In this study, four temporal satellite images of 10-yearly intervals (1980, 1990, 2000, and 2010) have been classified to determine the urban extent and built-up growth of Kolkata. These digitally classified imageries then have been used for the analysis. The analysis has been performed in consideration of jurisdictional boundary of Kolkata Municipal Corporation as well as the natural boundaries of the city. The main problem faced in this analysis was lack of temporal ancillary datasets that were essential for many of the widely practiced metrics. Therefore, the intention was to use some metrics that are less demanding in terms of data and computation. However, it has been found that most of these metrics are inferior in capturing insights into urban growth and sprawl. In this study, therefore, most of thcsc mctrics have been modified, either to overcome their limitations or to fit the study area.

Although the research documented in this book aims to analyze the urban growth of Kolkata, the book may be referred by urban researchers worldwide, especially those using remote sensing data for their analysis. They will benefit from the critical discussions and demonstrated methods/models.

This book comprises five chapters and three appendices. Each chapter starts with a brief of the topics to be covered within the chapter. Chapter 1 provides an introduction and overview. It focuses on the background, significance, objectives, and methodological overview of the research. Chapter 2 discusses the review of literature. This chapter briefly documents the patterns and processes of urban growth and sprawl, and application of remote sensing data in such analysis. It also critically reviews the analytical and quantification techniques of urban growth and sprawl from remote sensing data. Finally, it progresses toward the scope of the research. Chapter 3 is intended to describe the data and methodology. It describes the data and software that have been used; how the remote sensing data have been rectified and classified; how the accuracy assessment has been performed; what were the steps involved in preparing the vector maps of Kolkata; how the census data have been encoded; and how the built-up data have been extracted. Analytical steps have also been explained in this chapter. Critical comments on the analytical techniques, their evaluations, and justifications on reliance have also been discussed. Chapter 4 deals with the results obtained from the analysis. This chapter presents the built-up data in terms of classified images and also of matrices. It shows the status of urban growth and sprawl for the city of Kolkata. This chapter also aims to discuss the findings, make arguments on the methodology in terms of their merits and limitations, and construct logical statements. Chapter 5 draws the conclusions and highlights the future scope of research and study. Appendix A briefly describes the study area, its geographical location and properties, administrative and urban structures. Appendix B furnishes the census data for each municipal ward and borough in Kolkata Municipal Corporation. Appendix C furnishes the derived built-up data for each borough of Kolkata Municipal Corporation, as well as for each natural boundary of the city.

Acknowledgments

I am grateful to all the authors of the numerous books and research publications mentioned in the list of references of this book. These valuable literatures formed the basis of my knowledge required for conducting the research and writing this title. I express my gratitude to those teachers, researchers, and organizations for their contributions that reinforced my knowledge.

I would like to express my profound gratitude to Dr. Subhajit Saraswati, Professor, Department of Construction Engineering, Jadavpur University, and Dr. Debasish Bandyopadhyay, Associate Professor, Department of Construction Engineering, Jadavpur University, Kolkata. Their guidance was an invaluable resource for this research. I am much grateful to Prof. Rana Dattagupta, Director, CAD Centre, Jadavpur University, for providing all possible facilities to execute this research and prepare this title.

I am thankful to my colleagues, especially Mr. Biswajit Giri, Mr. Chiranjib Karmakar, Mr. Subrata Das, Mr. Santanu Glosal, and Mr. Uday Kumar De. Without their help and cooperation executing this research was never possible.

Finally I express my gratitude to my parents who have been a perennial source of inspiration and hope for me. I also want to thank my wife Chandrani, for her understanding and full support, while I worked on this research. My little daughter, Bagmi, deserves a pat for bearing with me during this rigorous exercise.

Basudeb Bhatta

Contents

Chapter 1
Introduction

Abstract This chapter introduces the study and research focused in this book. It aims to describe the background, significance, and objectives of the research; overview, questions that are to be answered in the research, and methodological overview. This chapter, as expected from any research-based book, acts as a prelude to all other chapters. The background section briefs the urban growth, sprawl, and the importance of remote sensing data and GIS techniques in urban growth analysis. The significance section documents the importance of current research. Other three sections (research objectives, research questions and methodological overview) will help to shape the readers' outlook on the research.

Keywords Urban growth · Sprawl growth · Kolkata · India · Remote sensing · GIS

1.1 Background

A spatial and demographic process that concentrates human population within a particular economy and society is generally regarded as urban growth. People move from villages to towns and cities in expectations of better life and economic opportunities. This movement is the primary cause of urban growth. A direct consequence of urban growth can be observed as increase in built areas.

Urban sprawl is an undesirable pattern as well as process of urban growth. This type of growth generally extends the core urban area to the outside rapidly in a dispersed manner. It occurs during the suburbanization process of residence, industry, and commerce. Sprawl encroach large amount of farmland, open space, and forest; may bring negative impacts on environment and may cause more traffic problems.

B. Bhatta, *Urban Growth Analysis and Remote Sensing*, SpringerBriefs in Geography,
DOI: 10.1007/978-94-007-4698-5_1, © The Author(s) 2012

The concept of sprawl-emergence of a situation of unauthorized and unplanned development, normally at the fringe areas of cities, especially haphazard and piecemeal construction of homesteads, commercial areas, industrial areas, and other non-conforming land-uses, generally along the major lines of communications or roads adjacent to specified city limits, is observed which is often termed as the urban sprawl. The area of urban sprawl is characterized by a situation where urban development adversely interferes with urban environment which is neither an acceptable urban situation nor suitable for an agricultural rural environment (Rahman et al. 2008).

"Urban growth and sprawl is a pertinent topic for analysis and assessment today. The environmental impacts of urban sprawl and the extent of urban problems have been growing in complexity and relevance, generating strong imbalances between the city and its hinterland. The need to address this complexity in assessing and monitoring urban planning and management processes and practices is strongly felt" (EEA 2002).

In the industrialized countries the future growth of urban populations will be comparatively modest since their population growth rates are low and over 80 % of their population already live in urban areas. Conversely, developing countries, like India, are in the middle of the transition process, when urban population growth rates are very high. According to the United Nations report (UNFPA 2007), the number and proportion of urban dwellers will continue to rise quickly. Urban global population will grow to 4.9 billion by 2030. In comparison, the world's rural population is expected to decrease by some 28 million between 2005 and 2030. At the global level, all future population growth will thus be in towns and cities; most of which will be in developing countries. The urban population of Africa and Asia is expected to be doubled between 2000 and 2030 (UNFPA 2007). This huge growth in urban population, coupled with several other factors, may force uncontrolled urban growth. One of the biggest challenges for science, engineering, and technology in the twenty-first century is how to control the urban growth to protect natural resources and ecosystem.

The international participants are increasingly engaged with the urgent environmental tasks for the sustainable development of their urban regions, the planning challenges faced by the local authorities, and the application of remote sensing data and geographic information system (GIS) techniques for the analysis of urban growth to meet these challenges. Determining the rate of urban growth and urban spatial configuration, from remote sensing data, is a prevalent approach in contemporary urban geographic studies. Maps of growth and a classified urban structure derived from remotely sensed data can assist planners to visualize the trajectories of their cities, their underlying systems, functions, and structures. There are currently a number of applications of analytical methods and models available to cities by using the remote sensing data and GIS techniques, especially for mapping, monitoring, measuring, analyzing, and modeling (Bhatta 2010).

However, despite the promise of new and fast-developing remote sensing technologies, a gap exists between the research-focused results offered by the urban remote sensing community and the application of these data and methods/

models by the governments of urban regions. This condition is true especially in developing countries. There is no end to the interesting scientific questions to ask about cities and their growth, but sometimes these questions do not match the operational problems and concerns of a given city. This necessitates more focused research and debate in the areas of urban growth analysis, in view of their applications. The current research concentrates at the problems associated with the analysis of urban growth and sprawl, for the city of Kolkata, by using remote sensing data from a technological perspective, rather than scientific.

1.2 Significance

Urban development is the process of emergence of the world dominated by cities and by urban values (Clark 1982). The occurrence of urban development is so general, and its implications are so wide, that it is possible to view much of recent social and economic history in terms of the attempts to cope with its varying consequences. The rise of great cities and their growing spatial influence initiated a change from largely rural to predominantly urban places and patterns of living that has affected most countries over the last two centuries. Currently, not only do large numbers of people live in or immediately adjacent to towns and cities, but whole segments of the population are completely dominated by urban values, expectations and life styles (Clark 1982). From its origins as a locus of non-agricultural employment, the city has become the major social, cultural and intellectual stimulus in modern urban society. Urban growth is one of the forms of urban development; which in recent decades became the paramount concerns to politicians, economists, administrators, planners and developers, sociologists, environmentalists, and proponents from many other disciplines.

Determining the rate of urban growth and urban spatial configuration, from remote sensing data, is not a new topic in contemporary urban geographic studies, rather it has a long history. There are currently many applications of analytical and simulation models to cities by using the remote sensing data and GIS techniques (Batty 2000).

Analysis of urban growth and sprawl also attracted a wide research focus in the recent years, especially, by using remote sensing data. Many metrics and statistics have been proposed by the researchers in the last two decades that have especially been applied on cities in developed countries. Most of these metrics can measure different properties of urban growth in relative scales; they are inferior in capturing the urban growth and sprawl in absolute scales that can measure and evaluate the urban growth and sprawl in black-and-white characterization. Researchers, in general, used many metrics and scales toward the understanding of urban growth and sprawl. Owing to several scales and parameters used in these studies it is often difficult to get an overall idea from these measures (Bhatta et al. 2010a). Furthermore, although these models can provide a better insight into urban development, but in many instances they require extensive inputs in order to obtain the final result. In the industrialized countries they may have proper planning policies for their cities, however the cities in developing countries lack such type of policies in most of the cases and they grow with all freedoms. Many of the aforementioned models, which have been developed and appreciated in the international community, are mainly focused on the cities in industrialized countries. Therefore, the merits of such models for the cities in developing countries seem far less convincing. Furthermore, in several instances, the city administrators in developing countries are not well conversant with the new tools or methods and modern technologies such as geospatial technology. Therefore, they demand simple analytical approaches that require minimal set of input data.

It has also been noticed that although many metrics and statistics are being used in the existing practices of urban growth analysis, there is no such quantifiable scale that can determine the goodness of urban growth in a single measure. In conclusion, whether the growth in a specific urban area is good or bad, is often based on the intuition of the analyst. It would be better if mathematical models could help us in doing so. The concept of goodness in urban growth was earlier introduced by the author and his colleagues (Bhatta et al. 2010b). But, the method had several limitations. This title proposes a modified approach to overcome those limitations.

The city of Kolkata, the capital of the Indian state of West Bengal, is one of the largest cities in India. It is the main business, commercial and financial hub of eastern India and the northeastern states. The city is more than 300 years old and it served as the capital of India during the British governance until 1911. However, despite of being one of the oldest cities in West Bengal and deserving the highest commercial and political importance, the city has never experienced properly designed planning policies as a whole. Several unplanned and patchy developmental initiatives by the local governments have resulted in serious problems viz. disparity in distribution of urban facilities and wealth, sprawl, traffic congestion, and pollution. Therefore, there was an urgent need to analyze the city's urban growth for the past and present in order to prepare for the future in view of the sustainable development; although till this research initiated, no such significant attempt could be seen. This research was initiated in the year 2006 with the objectives listed in Sect. 1.3. During the research, several existing methods and models have been tested; new methods and models have also been developed. The intermediate outcomes of the research have been reported in several journals (Bhatta 2009a, b; Bhatta et al. 2010b). However, as the research progressed, it has been found that in many instances the reported methods and models have several limitations. Later, either they have been replaced with a new one or modified to overcome their limitations. This title documents the final outcomes of the research as a whole.

1.3 Objectives

The main objective of this research was to analyze the urban growth and sprawl of Kolkata for the last three decades (1980–2010) by using remote sensing data. Since the city of Kolkata lacks multi-temporal ancillary datasets for the past which were essential in the conventional ways of analysis, the research had to find out some techniques that can suite this study area. In doing so, the research objectives have been diversified as the evaluation of some existing models to the modification of them, and to propose new models.

The objectives of this research were, therefore (1) to analyze the status of urban growth in Kolkata for the past and present; (2) to evaluate some simple but effective analytical approaches with a minimum set of input data; (3) to identify the merits and limitations of some neoclassical models for the analysis of urban growth and sprawl for the considered study area; (4) to modify some neoclassical models to overcome their limitations that would be justified by the current research; and last but not least (5) to propose new mathematical and schematic models for the analysis of urban growth and sprawl which may be suitable for the cities in developing countries as well as worldwide.

The emphasis, in this research, was on black-and-white characterization of urban growth and sprawl, rather than relative measures.

1.4 Research Questions

The objectives, as stated in the previous section, form the following two sets of research questions:

(1) What was the status of urban growth in the city of Kolkata in last three decades? Whether the growth was increasing or decreasing? Did the urban growth meet the growth rate of population within this area? How the per capita consumption of built-up area changed with time? Was the urban growth compact or dispersed? What was the degree of compactness or dispersion? Whether the observed growth is acceptable or not?
(2) The aforementioned questions form another set of research questions, such as: What may be the preferred techniques to extract the information from the remote sensing imageries? To what extent the data extracted from remote sensing imageries are reliable? Is it possible to conduct such research without the support of remote sensing data? What are the metrics and models available to analyze the urban growth from the remote sensing imageries? What may be the merits and demerits of those metrics and models? Are they applicable universally, especially in case of Kolkata? If yes, then, do they reveal satisfactory results in case of Kolkata? Can they be justified through logical or mathematical evaluations? Is there any necessity of modifying these models

for more reliable or useful results? How can they be modified? Can a new model be introduced for achieving better results?

It has been assumed that if these questions can be answered through a serious study, the research will achieve its goals eventually.

1.5 Methodological Overview

The methodology started with the processing of remote sensing data toward the determination of urban area by means of digital classification. These classified imageries determined the urban growth rates and their related properties. The analyses have been divided into two distinct parts: (1) in consideration of the administrative boundary of Kolkata Municipal Corporation (KMC), and (2) in consideration of the natural boundary of the city.

In the first part, analyses were performed for the last three decades (1980–2010) to understand the pattern and process of urban growth and sprawl in the KMC area. Four multispectral remote sensing imageries in ten-yearly intervals (1980, 1990, 2000, and 2010) have been used to conduct the research. Census data have also been used to evaluate the urban growth in relation to populations and number of housing units within the KMC. Several metrics have been tested, modified, and proposed in this part of analysis.

In the second part, the natural boundaries of the city, for four temporal dates (1980, 1990, 2000, and 2010), have been determined by means of digital neighborhood search on the classified imageries. Then, some other metrics (such as, area index, shape index, entropy) have also been applied to quantify the sprawl. Efforts have also been made to make some adjustments within these metrics to overcome their limitations. To check the degree of freedom (the extent to which the observed growth meets the planned growth), Chi-square statistic has been applied in a modified approach. Finally, to measure the goodness of urban growth, a modified approach has been proposed.

References

Batty M (2000) Geo-computation using cellular automata. In: Openshaw S, Abrahart RJ (eds) GeoComputation. Taylor & Francis, New York, pp 95–126

Bhatta B (2009a) Analysis of urban growth pattern using remote sensing and GIS : a case study of Kolkata, India. Int J Remote Sensing 30(18):4733–4746

Bhatta B (2009b) Modelling of urban growth boundary using geoinformatics. Int J Digit Earth 2(4):359–381

Bhatta B (2010) Analysis of urban growth and sprawl from remote sensing data. Springer-Verlag, Heidelberg, p 172

Bhatta B, Saraswati S, Bandyopadhyay D (2010a) Urban Sprawl Measurement from Remote Sensing Data. Appl Geography 30(4):731–740

Bhatta B, Saraswati S, Bandyopadhyay D (2010b) Quantifying the degree-of-freedom, degree-of-sprawl, and degree-of-goodness of urban growth from remote sensing data. Appl Geography 30(1):96–111

Clark D (1982) Urban geography: an introductory guide. Taylor & Francis, London, p 231

EEA (European environment agency) (2002). Towards an urban atlas: assessment of spatial data on 25 European cities and urban areas. EEA, Copenhagen

Rahman G, Alam D, Islam S (2008) City growth with urban sprawl and problems of management. In Proceedings of 44th ISOCARP Congress Dalian, China, Sept 19–23, International Society of City and Regional Planners and Urban Planning Society of China

UNFPA (United nations population fund) (2007). Peering into the dawn of an urban millennium, State of world population 2007: Unleashing the Potential of Urban Growth. URL: www.unfpa.org/swp/2007/english/introduction.html

Chapter 2
Review of Literature

Abstract The literature on urban growth and sprawl is bigger than the scope of this chapter. Application of remote sensing and GIS in the study of urban growth/ sprawl is also extensive in terms of their variations. A general research review cannot embrace the totality. However, in this chapter, efforts have been made to briefly document the important aspects that mainly focus on urban growth and sprawl, their patterns and processes, and measurement/analysis. It has been found that sprawl as a concept suffers from difficulties in definition. As a result, quantification of this phenomenon is rather complicated and sometimes confusing, especially if we use remote sensing data. There are dozens of metrics that are practiced by the urban planners and administrators in their cities, especially in developed countries. Merits and demerits of these measurements/analytical techniques have also been addressed briefly. It also documents similar studies in India, and proceeds toward the proposed research by documenting the scope of research.

Keywords Urban growth · Sprawl measurement · Remote sensing · GIS · Sprawl technique advantages

2.1 Urban Growth and Sprawl

The spatial configuration and the dynamics of urban growth are important topics of analysis in the contemporary urban studies. Several studies have addressed these issues which have dealt with diverse range of themes (e.g., Acioly and Davidson 1996; Wang et al. 2003; Páez and Scott 2004; Zhu et al. 2006; Hedblom and Soderstrom 2008; Geymen and Baz 2008).

Urban sprawl, as a concept, suffers from difficulties in definition (Johnson 2001; Barnes et al. 2001; Wilson et al. 2003; Roca et al. 2004; Sudhira and Ramachandra 2007; Angel et al. 2007; Bhatta 2010). Galster et al. (2001) critiqued the

B. Bhatta, *Urban Growth Analysis and Remote Sensing*, SpringerBriefs in Geography,
DOI: 10.1007/978-94-007-4698-5_2, © The Author(s) 2012

conceptual ambiguity of sprawl observing that much of the existing literature is 'lost in a semantic wilderness.' Their review of the literature found that sprawl can alternatively or simultaneously refer to: (1) certain patterns of land use, (2) processes of land development, (3) causes of particular land-use behaviors, and (4) consequences of land-use behaviors. They have reviewed many definitions of sprawl from different perspectives. It seems that sprawl is used both as a *noun* (condition) and as a *verb* (process), and suffers from lack of clarity even though many would claim to 'know it when they see it' (e.g., Ewing 1994).

Although accurate definition of urban sprawl is debated, a general consensus is that urban sprawl is characterized by unplanned and uneven pattern of growth, driven by multitude of processes and leading to inefficient resource utilization. The direct implication of such sprawl is change in land-use and land-cover of the region as sprawl induces the increase in built-up and paved area (Sudhira and Ramachandra 2007). It is worth mentioning that opinions on sprawl held by researchers, policy makers, activists, and the public differ sharply, and the lack of agreement over how to define sprawl certainly complicates the efforts to characterize and restrict this type of land development. Researchers have made many attempts to characterize the sprawl as shown in Table 2.1.

Table 2.1 clearly shows that sprawl can be analyzed from many perspectives and it may vary from case to case. However, most of the researchers have emphasized the density (e.g., built-up density, population density, and housing density). Therefore, it can be considered that the best characterization of sprawl is considering the density criterion.

2.2 Physical Patterns and Forms of Urban Growth and Sprawl

Wilson et al. (2003) identified three categories of urban growth: *infill*, *expansion*, and *outlying*; with outlying urban growth further separated into *isolated*, *linear branch*, and *clustered branch* growth (Fig. 2.1). The relation (or distance) to existing developed areas is important when determining what kind of urban growth has occurred.

Indeed all types of urban growth are not considered as sprawl; one development that can be considered as sprawl by someone may not be considered by others (Roca et al. 2004). Furthermore, urban sprawl has a negative connotation, and not all urban growth is necessarily unhealthy. In fact, some types of urban growth (e.g., infill growth) are generally considered as remedies to sprawl. Therefore, sprawl cannot be characterized by the simple quantification of 'the amount of land that has changed to urban uses.' Sprawl phenomenon should be treated separately than the general urban growth. Ewing (1994) reviewed several patterns of urban sprawl, and argued that the pattern of sprawl is 'like obscenity', the experts may know sprawl when they see it.

Table 2.1 Varying characterization of sprawl (after Torrens 2008)

	Growth	Social	Aesthetic	Decentralization	Accessibility	Density	Open space	Dynamics	Costs	Benefits
Audirac et al. (1990)		●								
Bae and Richardson (1994)						●				●
Benfield et al. (1999)									●	
Burchell et al. (1998)	●		●		●	●			●	
Calthorpe et al. (2001)			●							
Clapham (2003)							●			
Duany et al. (2001)			●							
El Nasser and Overburg (2001)						●				
Ewing (1997)		●		●	●	●	●	●	●	
Ewing et al. (2002)		●		●	●	●	●	●	●	
Farley and Frey (1994)		●								
Galster (1991)		●		●						
Galster et al. (2001)	●			●						
Gordon and Richardson (1997a)						●		●		●
Gordon and Richardson (1997b)						●		●		●
Hasse and Lathrop (2003a)							●			
Hasse and Lathrop (2003b)			●	●	●					
Hasse (2004)					●					
HUD (1999)					●					
Lang (2003)						●				
Ledermann (1967)						●				
Lessinger (1962)				●						
Malpezzi (1999)					●					
OTA (1995)				●						
Peiser (1989)						●				
Pendall (1999)						●				
RERC (1974)						●			●	
Sierra Club (1998)					●					
Sudhira et al. (2004)							●			

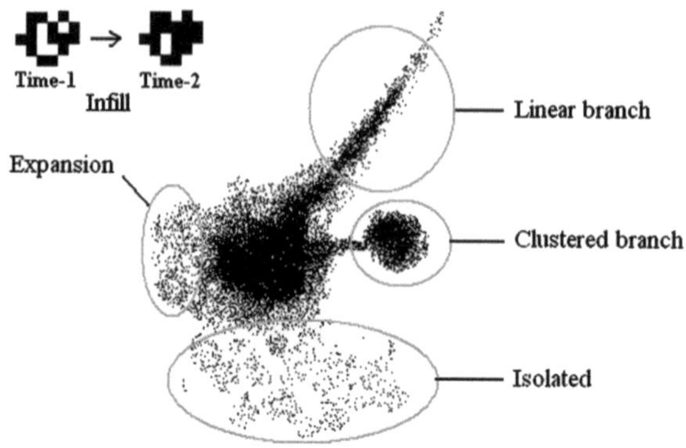

Fig. 2.1 Schematic diagram of urban growth pattern (Bhatta 2010)

2.3 Temporal Process of Urban Growth and Sprawl

Galster et al. (2001) considered the urban sprawl both as a *pattern* of urban land use (a spatial configuration of a metropolitan area at a specific time); and as a *process*, namely as the change in the spatial structure of cities over time. Sprawl as a pattern or a process is to be distinguished from the causes that bring such a pattern about, or from the consequences of such patterns (Galster et al. 2001). If the sprawl is considered as a pattern it is a static phenomenon and as a process the sprawl is a dynamic phenomenon. Some of the researchers have considered sprawl as a static phenomenon, whereas some have analyzed it as a dynamic phenomenon; however, most of the researchers shout for both.

Sprawl, as a pattern, helps us to understand the spatial distribution but as a static phenomenon, in fact areas described as sprawled are typically part of a dynamic urban scene (Harvey and Clark 1965; Ewing 1997). The dynamics of sprawl process can be understood from the theoretical framework of urban growth process. Herold et al. (2005b) present a hypothetical schema of urban growth process using a general conceptual representation as shown in Fig. 2.2. According to them, urban area expansion starts with a historical seed or *core* that grows and disperses to new individual development centers. This process of *diffusion* continues along a trajectory of organic growth and outward expansion. The continued spatial evolution transitions to the *coalescence* of the individual urban blobs. This phase transition initially includes development in the open space in interstices between the central urban core and peripheral centers. This conceptual growth pattern continues and the system progresses toward a saturated state. In Fig. 2.2, this 'final' agglomeration can be seen as an initial urban core for further urbanization at a less detailed zoomed-out extent. In most traditional urbanization-studies this 'scaling up' has been represented by changing the spatial extent of concentric rings around the central urban core.

The preceding framework suggests that some parts of an urban area may pass through a sprawl stage before eventually thickening so that they can no longer be characterized as sprawl. However, from this point of view, what, when, and where

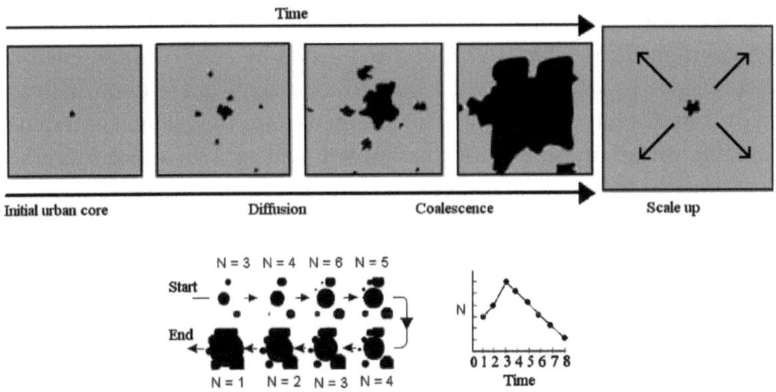

Fig. 2.2 Sequential frames of urban growth. The graph on the *bottom-right* shows *N*, number of agglomerations, through a sequence of time steps (Herold et al. 2005b)

it can be characterized as a sprawl becomes complex. Therefore, sprawl as a process without considering the pattern cannot be characterized. Rather, it should be considered as a pattern at multiple dates. "In any event, measuring the respective dimensions of development patterns for an urban area at different times will reveal the process (or progress) of sprawl" (Galster et al. 2001).

One may refer Bhatta (2010) for detailed discussion on urban growth pattern and process. The title also addresses the causes and consequences of urban growth and sprawl, smart and sustainable urban growth, and urban planning policies to restrict the sprawl. The title cites several references that may be useful for the study of urban growth/sprawl. Since the primary goal of current research is not focused on urban planning, rather focused on the analysis of urban growth and sprawl from remote sensing data, the documentation of review on urban growth/sprawl/planning has been kept limited.

2.4 Remote Sensing for Analysis of Urban Growth and Sprawl

Understanding the urban patterns, dynamic processes, and their relationships is a primary objective in the urban research agenda with a wide consensus among scientists, resource managers, and planners, because future development and management of urban areas require detailed information about ongoing processes and patterns.

Remote sensing, although challenged by the spatial and spectral heterogeneity of urban environments (Jensen and Cowen 1999; Herold et al. 2004), seems to be an appropriate source of urban data to support such studies (Donnay et al. 2001). It is irrefutable that the *earth observation* is a modern science, which studies the earth's changing environment, through remote sensing tools such as satellite imagery and aerial photographs (EEA 2002). A report published by NASA highlighted the fact that the advances in satellite-based land surface mapping are contributing to the creation of considerably more detailed urban maps, offering planners a much deeper understanding of the dynamics of urban growth and sprawl, as well as associated matters relating to territorial management (NASA 2001).

In terms of analyzing urban growth, Batty and Howes (2001) believe that remote sensing technology, especially considering the recent improvements, can provide a unique perspective on growth and land-use change processes. Data sets obtained through remote sensing are consistent over great areas and over time, and can provide information at a great variety of geographic scales. The information derived from remote sensing can help to describe and model the urban environment, leading to an improved understanding that benefits applied urban planning and management (Banister et al. 1997; Longley and Mesev 2000; Longley et al. 2001).

Analysis of urban growth from remote sensing data, as a pattern and process, helps us to understand how an urban landscape is changing through time. This understanding includes: (1) the rate of urban growth, (2) the spatial configuration of growth, (3) whether there is any discrepancy in the observed and expected growth, (4) whether there is any spatial or temporal disparity in growth, and (5) whether the growth is sprawling or not.

In the recent years, remote sensing data and geographic information system (GIS) techniques are widely being used for mapping (to understand the urban pattern), monitoring (to understand the urban process), measuring (to analyze), and modeling (to simulate) the urban growth, land-use/land-cover change, and sprawl. The physical expressions and patterns of urban growth and sprawl on landscapes can be detected, mapped, and analyzed by using remote sensing data and GIS techniques. The decision support systems within the GIS can evaluate remote sensing and other geospatial datasets by using multi-agent evaluation (Axtell and Epstein 1994; Parker et al. 2003) which can also predict the possibilities in the subsequent years using the current and historical data. In the last few decades, these techniques have successfully been implemented to detect, analyze, and model the urban growth dynamics. All these issues have been reviewed and documented with a greater detail in Bhatta (2010).

2.5 Measurement and Analysis of Urban Growth and Sprawl

The process of mapping urban growth results in the creation of abstracted and highly simplified change maps of the study area (Singh 1989). Recently, urban change detection focus has shifted from detection to quantification of change, measurement of pattern, and analysis of pattern and process of urban growth and sprawl.

Urban growth can be quantified by measuring the built-up change between two dates (Singh 1989; Jensen 2005). However, quantification of sprawl, as a pattern or process, is a real challenging issue (Bhatta et al. 2010a). Wilson et al. (2003) said that without a universal definition, quantification and modeling of urban sprawl is extremely difficult. They argue that creating an urban growth model instead of an urban sprawl model allows us to quantify the amount of land that has changed to urban uses, and lets the user decide what he or she considers as urban sprawl. Angel et al. (2007) also support this concept. This statement, however, discourages the efforts of quantifying the sprawl and makes the sprawl phenomenon more ambiguous. "Although there have been many studies on the measurement of urban form they have limitations in capturing the characteristics of urban sprawl" (Yeh and Li 2001); the results obtained from such measurement processes are often not easily interpretable. Several limitations of remote sensing data have made the quantification processes more difficult (Du et al. 2002; Prenzel 2004; Paolini et al. 2006; Hardin et al. 2007).

Sprawl can be measured in relative and absolute scales (Bhatta et al. 2010a). Absolute measurements are capable to create a black-and-white distinction between a sprawled city and a compact city. Relative measures, in contrast, quantify several attributes that can be compared among cities, among different zones within a city, or among different time (date) for a city. In case of relative measures, whether the city is sprawled or not is generally decided by the analyst, or even left without characterizing the sprawl. It is important to mention that most of the sprawl measurement techniques, in general, are relative measures, and can be used as indicators of sprawl. Absolute measurement of sprawl is never possible with these measures, unless we define a threshold toward the black-and-white characterization of sprawling and nonsprawling. Defining a threshold is not an easy task. Researchers have made their own assumptions toward defining this threshold, which are even less clear to the scientists (Bhatta et al. 2010a). It is important to realize that relative measures, most often, fail to draw conclusion on sprawl, and a threshold used in one area cannot be used in other areas reliably. These measures may serve the scientific purposes well, but can never become a technology, because to interpret the results one has to be a scientist. Therefore, how these techniques can become a tool for a city administrator is an obvious question.

Many metrics and statistics have been used to measure the sprawl. These metrics are generally known as *spatial metrics*. Spatial metrics are numeric measurements that quantify spatial patterning of land-cover patches, land-cover classes, or entire landscape mosaics of a geographic area (McGarigal and Marks 1995). These metrics have long been used in *landscape ecology* [where they are known as *landscape metrics* (Gustafson 1998; Turner et al. 2001)] to describe the ecologically important relationships such as connectivity and adjacency of habitat reservoirs. Applied to research fields outside of landscape ecology and across different kinds of environments (in particular, urban areas), the approaches and assumptions of landscape metrics may be more generally referred to as *spatial metrics* (Herold et al. 2005a). Spatial or landscape metrics, in general, can be defined as quantitative indices to describe structures and patterns of a landscape (O'Neill et al. 1988). Herold et al. (2005a) defined it as 'measurements derived from the digital analysis of thematic-categorical maps exhibiting spatial heterogeneity at a specific scale and resolution.'

Spatial metrics have found important applications in quantifying urban growth, sprawl, and fragmentation (Hardin et al. 2007). Based on the work of O'Neill et al. (1988), sets of different spatial metrics have been developed, modified, and tested (Hargis et al. 1998; McGarigal et al. 2002; Riitters et al. 1995). Many of these quantitative measures have been implemented in the public domain statistical package FRAGSTATS (McGarigal and Marks 1995; McGarigal et al. 2002).

Spatial metrics can be grouped into three broad categories: *patch*, *class*, and *landscape* metrics. *Patch metrics* are computed for every patch in the landscape, *class metrics* are computed for every class in the landscape, and *landscape metrics* are computed for entire patch mosaic. There are numerous types of spatial metrics that are found in the existing literature, for example: area/density/edge metrics (patch area, patch perimeter, class area, number of patches, patch density, total edge, edge density, landscape shape index, largest patch index, patch area distribution); shape metrics (perimeter-area ratio, shape index, fractal dimension index, linearity index, perimeter-area fractal dimension); core area metrics (core area, number of core areas, core area index, number of disjunct core areas, disjunct core area density, core area distribution); isolation/proximity metrics (proximity index, similarity index, proximity index distribution, similarity index distribution); contrast metrics (edge contrast index, contrast-weighted edge density, total edge contrast index, edge contrast index distribution); contagion/interspersion metrics (percentage of like adjacencies, clumpiness index, aggregation index, interspersion and juxtaposition index, mass fractal dimension, landscape division index, splitting index, effective mesh size); connectivity metrics (patch cohesion index, connectance index, traversability index); diversity metrics (patch richness, patch richness density, relative patch richness, Shannon's diversity index, Simpson's diversity index, Shannon's evenness index, Simpson's evenness index); and many others (McGarigal and Marks 1995; McGarigal et al. 2002).

The question is which spatial metrics are most appropriate for the measurement and analysis of sprawl. Galster et al. (2001) identified eight conceptual dimensions of land-use patterns for measuring the sprawl. Under the name of *sprawl metrics*, Angel et al. (2007) have demonstrated five metrics for measuring manifestations of sprawl and five attributes for characterizing the sprawl. Under each attribute they have used several metrics to measure the sprawl phenomenon. However, they had not recommended any threshold that could be used for distinguishing a sprawling city from a non-sprawling city. Furthermore, interpretation of results from these metrics is also difficult and confusing since metrics are huge in number and one may contradict with other. Refer Bhatta et al. (2010a) for more detailed discussion.

Sierra Club (1998) ranked major metropolitans in USA by four metrics, including: (1) population moving from inner city to suburbs; (2) comparison of land use and population growth; (3) time cost on traffic; and (4) decrease of open space. USA Today (2001) puts forward the share of population beyond *standard metropolitan statistical area* as an indicator for measuring sprawl. Smart Growth

America (Ewing et al. 2002) carried out a research to study the impacts of sprawl on life quality in which four indices had been used to measure urban sprawl: (1) residential density; (2) mixture of residence, employment, and service facilities; (3) vitalization of inner city; and (4) accessibility of road network. All of these metrics were relative measures and failed to black-and-white discrimination of sprawling and nonsprawling.

Some of the researchers have also contributed to measuring sprawl by establishing multi-indices by GIS analysis or descriptive statistical analysis (Nelson 1999; Kline 2000; Torrens 2000; Galster et al. 2000, 2001; Hasse 2004). These indices cover various aspects including population, employment, traffic, resources consumption, architecture aesthetics, living quality, etc. Commonly used indices include: *growth rate* (of a population or built-up area); *density* (population density, residential density, employment density); *spatial configuration* (fragmentation, accessibility, proximity); and others (per-capita consumption of land, land-use efficiency etc.) (e.g., The Brookings Institution 2002; USEPA 2001; Fulton et al. 2001; Masek et al. 2000; Pendall 1999; Sutton 2003; Jiang et al. 2007). However, no one has provided any straight answer to the questions like: what should be the built-up growth rate in a non-sprawling city, or what should be the per-capita consumption of land in a non-sprawling city.

Torrens (2008) argues that sprawl should be measured and analyzed at multiple scales. In his (her) approach to measuring sprawl, (s)he has declared some ground-rules in developing the methodology. Measurements have been made to translate descriptive characteristics to quantitative form. The analysis is focused at micro-, meso-, and macro-scales and can operate over net and gross land. The analysis examines sprawl at city-scale and at intraurban levels—at the level of the metropolitan area as well as locally, down to the level of land parcels. Although interurban comparison and the use of remote sensing data are not focused on in this chapter, the methodology should be sufficient to be generalized to other cities using remote sensing data. The methodology devised a series of 42 measures of sprawl, which have been tracked longitudinally across a 10-year period. Although the author claims that this approach can provide a real insight of urban sprawl, however, the methodology became complex and resulted in confusion owing to the use of many scales and metrics.

Jiang et al. (2007) proposed 13 attributes under the name of 'geospatial indices' for measuring the sprawl in Beijing. Finally, they proposed an *integrated urban sprawl index* that combines the preceding 13 indices. This approach, indeed, minimizes the interpretation effort. However, their approach requires extensive inputs of multi-temporal data such as population, GDP, land-use maps, land-use master planning, floor-area ratio, maps of highways, and maps of city centers. Many developing countries lack such type of temporal data, and therefore, most of these indices are difficult to derive. Furthermore, they did not mention any threshold to characterize a city as sprawling or nonsprawling. However, this type of temporal analysis is useful to compare between cities or different zones within a city or status of a city at different dates. Whether a city is becoming more sprawling or not, with the change of time, can be well depicted by this type of analysis.

The main problem associated with most of the available sprawl measurement scales is the failure to define the threshold between sprawling and nonsprawling. Although relative comparisons can provide us some insights into sprawl phenomenon and the associated city, but often these measures are not adequate and we need black-and-white characterization of sprawl. The second greatest problem is the number of metrics used for the measurement of sprawl. The preceding discussion shows that many scales and parameters are being used for the measurement of sprawl. The question is what the most stringent tools are or how effective they are. The answer is still awaited. Geoghegan et al. (1997), Alberti and Waddell (2000), Parker et al. (2001), and Herold et al. (2003) propose and compare a wide variety of different metrics for the analysis of urban growth. However, their comparisons do not suggest any standard set of metrics best suited for use in urban sprawl measurement as the significance of specific metric varies with the objective of the study and the characteristics of the urban landscape under investigation.

Important to mention, many metrics are correlated and thereby contain redundant information. Riitters et al. (1995) examined the correlations among 55 different spatial metrics by factor analysis and identified only five independent factors. Thus, many typical spatial metrics do not measure different qualities of spatial pattern. The analyst should select metrics that are relatively independent of one another, with each metric (or grouping of metrics) able to detect meaningful structure of urban landscape that can result in reliable measures of sprawl. It is often necessary to have more than one metric to characterize an urban landscape because one metric cannot say about all. However, the use of many metrics results in many measures which are often difficult to interpret resulting in difficulties for reaching to a black-and-white conclusion. Use of highly correlated metrics does not yield new information, rather makes interpretations more difficult. "Just because something can be computed does not mean that it should be computed" (Turner et al. 2001). Often, different metrics may also result in opposite conclusions; for example, in Herold et al. (2003), 'number of patches' within the time span 1929–1976 was increasing (an indication of sprawl); however, if one considers 'mean nearest neighbour distance between individual urban patches', it was decreasing (an indication of compactness).

Another challenge is the spatial resolution of remote sensing data. Many metrics, for example patch or spatial heterogeneity analysis, are dependent on spatial resolution. In a low spatial resolution image, individual objects may appear artificially compact or they may get merged together. In an area of low density development where houses are relatively far apart, a spatial resolution of 30 m will produce an estimate of developed land four times that produced using the same underlying data but a spatial resolution of 15 m. Apparently, the most preferred spatial data are those that are sufficiently fine scale to represent individual units, e.g., individual land parcels or houses. Indeed higher spatial resolution provides better interpretability by a human observer; but a very high resolution leads to a high object diversity which may end up in problems when a classification algorithm is applied to the data, or it may produce a very high number of patches resulting in complications in metric analysis. Due to the increased heterogeneity in high resolution images, analysis of spatial association or spatial heterogeneity will also be influenced at a high degree. Further, for multi-temporal analysis of sprawl or measurement of sprawl as a process may include images from sensors having different resolutions. In such cases, resolution-dependent metrics are no longer usable.

Furthermore, statistical properties and behaviors of some metrics are not well known (Turner et al. 2001). In cases where a single number is reported for a landscape, we may have little understanding of the degree to which landscape pattern must change to be able to detect significant change in the numeric value of the metric. Therefore, the analyst should definitely consider the criteria that will be applied to determine whether the obtained result from a metric is meaningful or not.

The density of built-up and intensity of annual growth can efficiently depict the sprawl features of low density and strong change, but they are still weak in capturing the particular spatial patterns of urban sprawl. These metrics are not usually spatially explicit; for example the growth of built-up area to the growth of households in a city. This type of metrics measures what is present and in their relative amounts, or properties, without reference to where on the landscape they may be located.

Entropy method, another urban sprawl metric, is perhaps the most widely used technique to measure the extent of urban sprawl with the integration of remote sensing and GIS (Yeh and Li 2001; Lata et al. 2001; Li and Yeh 2004; Sudhira et al. 2004; Kumar et al. 2007). Relative entropy can be used to scale the entropy value into a value that ranges from 0 to 1 (Thomas 1981). Yeh and Li (2001) argue that because entropy can be used to measure the distribution of a geographical phenomenon, the measurement of the difference of entropy between time t_1 and t_2 can be used to indicate the magnitude of change of urban sprawl.

Entropy is more robust, spatial, statistic than the others (Yeh and Li 2001; Bhatta et al. 2010a). Many studies have shown that entropy is better than the spatial dispersal statistics, such as the *Gini* and *Moran coefficients* (Tsai 2005), that are often dependent on the size, shape, and number of regions used in calculating the statistics, and the results can change substantially with different levels of areal aggregation (Smith 1975; Thomas 1981). This is a manifestation of the scale problem or *modifiable areal unit problem* (MAUP) (Openshaw 1984; Openshaw and Alvanidies 1999; Armhein 1995) which may exert unspecified influence on the results of spatial analysis (Openshaw 1991). The effects of the MAUP can be divided into two components: the *scale effect* and the *zoning effect* (Armhein 1995). The *scale effect* is the variation in numerical results that occurs due to the number of zones used in an analysis. The *zoning effect* is the variation in numerical results arising from different grouping systems of small areas into larger units. Figure 2.3a shows the change in means that occurs when smaller units are aggregated into larger units, and Fig. 2.3b shows the change in mean with the change of zoning system (shape and size).

Thomas (1981) indicates that relative entropy is better than traditional spatial dispersal statistics because its value is invariant with the value of n (number of

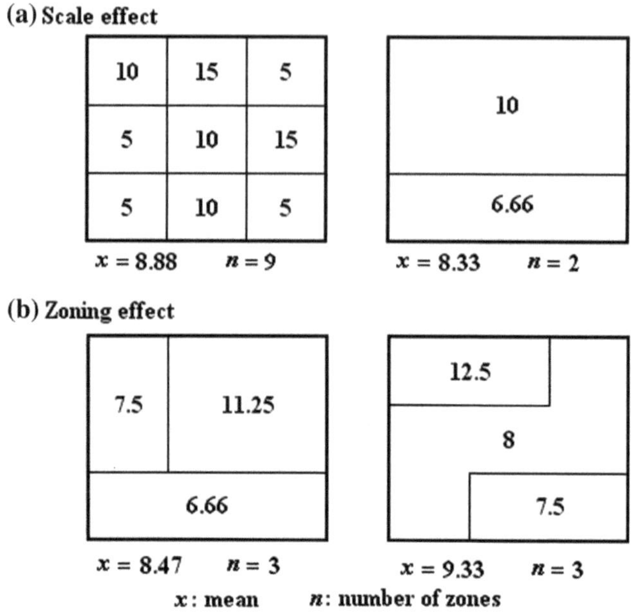

Fig. 2.3 The MAUP effects (Bhatta 2010). **a** Scale effect. **b** Zoning effect

zones). Therefore, the use of relative entropy can mitigate the scale effect of MAUP. However, relative entropy is still, to some extent, sensitive to the variations in the shapes and sizes of the regions used for collecting the observed proportions. For example, if there are two scales of analysis for the dispersion of population in a country, such as regions and subregions, different entropy values will be obtained if the data are collected based on regions instead of subregions. The entropy decomposition theorem can be used to identify different components of the entropy that are related to different zone sizes in collecting the data (Batty 1976; Thomas 1981; Yeh and Li 2001). It can alleviate the problem of comparing the results between different zone sizes because the influence of scaling can exactly be measured (Yeh and Li 2001).

Several other sprawl measurement techniques have also been reviewed and documented in Bhatta et al. (2010a) with a greater detail. However, it has been found that the entropy method is robust because of its preceding advantages.

In case of understanding the urban growth, analysis or quantification of sprawl is not adequate. In case of planned cities, the future growth is generally planned and modeled in advance. In many instances, the growth is restricted within a sharp defined boundary so that the actual growth does not exceed the planned growth. Therefore, whether the observed growth meets the planned (expected) growth and to what extent they meet are also necessary to be evaluated. Almeida et al. (2005) have used Chi-square statistics (degree of freedom) to evaluate the observed growth in comparison to expected growth. Their methodology, however, considers pattern and process in a single metric. Therefore, the urban growth pattern and process cannot be distinguished.

Since the entropy (degree of sprawl) and Chi-square (degree of freedom) are different measures and one may not relate other, it necessitates determining the 'degree of goodness' of the urban growth. The *degree of goodness* actually refers to the degree at which observed growth relates the planned growth and the magnitude of compactness (as opposed to sprawl). However, the review of literature did not find any metric that can quantify the goodness in urban growth in a single quantifiable scale. As stated in Chap. 1, the concept of goodness in urban growth was earlier introduced by the author and his colleagues during this research (Bhatta et al. 2010b). But, the method had several limitations (explained later in Sect. 3.14.5).

2.6 Administrative Versus Natural Boundary for the Analysis

Cities have administrative boundaries associated with them in the sense that city governments have jurisdiction over certain well-defined administrative areas. But the area contained within the jurisdictional boundary of a city has little to do with the metropolitan area of the city. In some cases, this area is very small in comparison with the size of the metropolitan area. The Los Angeles metropolitan area, for example, contains 35 independent municipalities. In other cases, for example in Beijing, the jurisdictional boundary of the municipality contains an area that is much larger than the built-up area of the city. The official area of the municipality is therefore not a very precise measure, neither of the built-up area of the city nor of what we intuitively grasp to be the city (Angel et al. 2005). Furthermore, the extent of a city is a dynamic phenomenon; it changes over time. However, the jurisdictional boundary of the city cannot be changed frequently owing to administrative complexities.

However, the main problem of the natural city extent is the lack of census data. Census data are generally collected and published in respect of census blocks; these blocks are defined in consideration of administrative boundary. Therefore, how the population or other socioeconomic parameters change over time within the natural boundary cannot be analyzed. If the analysis requires consideration of census observations then it must be based on the administrative tracts.

2.7 Determining the Natural Boundary of the City

Delineating the natural boundary or extent of a city is a real difficult task, because urban to rural transact shows a gradient (Fig. 2.4). If one interprets Fig. 2.4, he or she can understand the difficulty in demarcating the line that separates 'rural' and 'urban.'

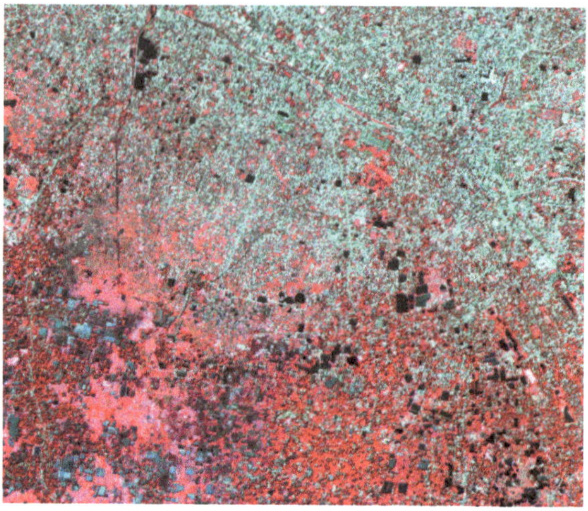

Fig. 2.4 Rural to urban gradient (Courtesy: National Remote Sensing Centre, India)

There are several approaches to determine the natural extent of a city or town (Bhatta 2010). Traditionally, the physical delimitation of urban areas and agglomerations has been characterized by two clearly differentiated approaches. On the one hand, delimitation is based upon physical or morphological criteria, where the continuous built-up area or the density of contiguous ambits comprises the basic mechanisms for the delimitation. On the other hand, studies based upon functional or economic criteria, the emphasis is placed upon the existing relations and flows throughout the urbanized territory, where the relation between place of residence and place of work is fundamental (Roca et al. 2004). Among these two, the former is more practiced. Population density is also another preferred option of defining urban area. In India, an area with a town or city of at least 50,000 people with continuous growth around it, encompassing a number of smaller towns and rural settlements based round the core town, with the possibility of being urbanized within the next couple of decades is defined as a standard urban area (Kundu and Basu 1999). One can understand the difficulty of delineating the urban area considering this definition. Furthermore, remote sensing data are incapable to convey such information. Therefore, in consideration of remote sensing data, physical or morphological criteria are the preferred approach to determine the natural extent of the city.

2.8 Subdivision of Natural City Extent

One of the major limitations of the several metrics (e.g., area index and shape index) is the consideration of the entire city in the analysis. They do not consider the variations of built-up areas in the different parts of the city. Therefore, a further analysis is essential to integrate the internal variations of the built-up area. To

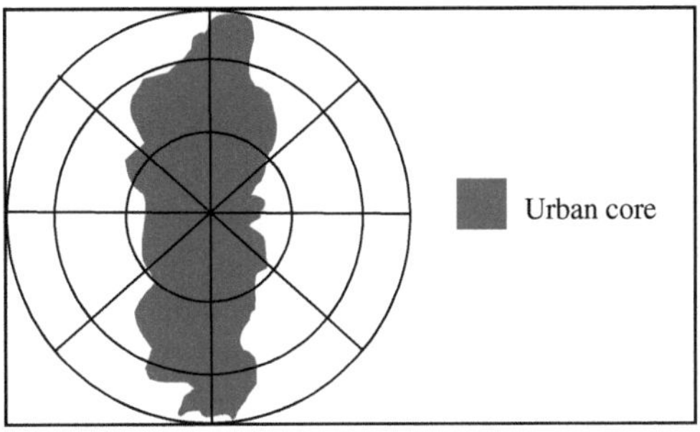

Fig. 2.5 A schematic diagram for the circular and pie-sectional subdivisions

understand the variations in built-up pattern, a popular approach is the moving window (Martinuzzi et al. 2007). It measures the spatial frequency[1]—pattern surrounding individual units of observation (i.e., a pixel) within the window. However, this measurement is highly dependent on the size of the window and there is no standard rule to define the window size. Further, it is influenced by the spatial resolution of the image data and can give varying results. For example, using remote sensing imagery with a spatial resolution of 60 m makes it more likely that small clusters of homogeneous classes will be classified as isolated cells; in comparison, measure from 30 m resolution image likely appears more scattered. On the other hand, aggregation (resampling) from 30 to 60 m resolution will also cause the pattern at the patch level to become simplified with fewer edges and a higher area-to-perimeter ratio, revealing a more compact urban pattern. Therefore, the area under analysis should be subdivided into smaller pieces independent of pixel array (window), and the aggregation (of built-up areas) should be done in consideration of these pieces.

There are different approaches to subdivide the study area. In case of analysis considering the administrative boundaries it is rather easier; administrative hierarchical boundaries may become the basis, for example, municipal boundary may be subdivided into boroughs or wards. In case of natural city extents, they are generally subdivided into circular or pie sections or in both (Kumar et al. 2007; Bhatta et al. 2010b). The main problem of this approach is its blindness to the actual shape of the city, for example, if the city is very long and narrow in shape the circular subdivision includes several rural (or other) areas in the analysis as shown in Fig. 2.5. Another better approach may be to construct buffered zones (positive or negative) by using the natural boundary of the city. However, for multitemporal analysis, we shall have multiple natural boundaries for multiple dates. Which boundary should be considered to construct the buffer is unclear. Therefore, we need a different approach for subdivision which can avoid the stated problems.

[1] Spatial frequency denotes the rate of changes in land-use/land-cover classes with the change of space.

2.9 Similar Studies in India

Although a chronological history of urban geographic research in India can be found in Thakur and Parai (1993), researches on urban growth, especially by using remote sensing data, have not been well documented. The first appearance of urban growth and sprawl analysis using remote sensing data in a publicly published form can be found in 1989 in the Journal of the Indian Society of Remote Sensing. Two papers, Sokhi et al. (1989) and Uttarwar and Sokhi (1989), were published; both were focused on the city of Delhi. Research of Uttarwar and Sokhi (1989) was aimed to study the urban fringe from aerial remote sensing data. This study was conducted for land-use change detection and inventory of land uses in suburban areas from Survey of India topographical maps of 1979, aerial photographs of 1984, and field check for updating in 1988. These data had been visually interpreted to generate land-use maps via manual mapping. Then these maps had been compared to quantify the urban growth; however, this chapter was not focused on sprawl. The other paper, Sokhi et al. (1989), was focused on mapping and monitoring of sprawl. Satellite imageries of 1975, 1981, 1985, and 1987 had been visually interpreted to prepare land-use/land-cover maps via manual mapping. Then the changes and built-up growth had been quantified from these maps. This chapter was on simple urban growth quantification. The distinction between urban growth and sprawl is not clear in this chapter.

Taragi and Pundir (1997) analyzed the urban growth and sprawl of Lucknow city for the period of 1972–1992. They derived nine major land-use/land-cover classes from four temporal remote sensing images via visual interpretation and manual mapping. They prepared a built-up change map and computed built-up growth rate. Based on these built-up information and land-use/land-cover maps, they tried to identify the sprawl intuitively. However, their 'sprawl-map' is actually a built-up change map and this research did not quantify the sprawl. Their characterization of sprawl was mainly limited to rapid growth.

A paper of Sudhira et al. (2003) was focused on the analysis of urban growth pattern in the form of either radial or linear sprawl along the Bangalore-Mysore highway over a period of 1972–1998. They used classified remote sensing image for 1998 and Survey of India topographical maps for the year 1972. Then the Shannon's entropy was used to determine the sprawl in different directions. Although this chapter successfully used the entropy method, however, consideration of only two temporal data set in a wide temporal gap is a questionable approach.

Sudhira et al. (2004) conducted a research to determine the dynamics of sprawl and to model the future sprawl using remote sensing and other dataset. They considered a study area of 434.2 km^2 in Mangalore–Udupi region in Karnataka state and the time span was considered 1972–1999. To determine the built-up information, they classified satellite remote sensing data for the year 1999. Built-up information of 1972 had been extracted from Survey of India topographical maps. Then they had analyzed the sprawl via built-up growth versus population growth, Shannon's entropy, pachiness, built-up density, population density, annual population growth rate, distance from Mangalore, and distance from Udupi. Metrics analysis had been performed using a 3 × 3 kernel window. Finally, they had predicted the built-up area for the years 2020 and 2050 by using regression analysis. However, although the title claimed for the modeling of sprawl, ultimately it resulted in a sum increase of built-up area for future.

Sudhira and Ramachandra (2007) conducted a similar analysis as the aforementioned one. They analyzed the urban growth and sprawl of Bangalore city using two temporal satellite remote sensing images for the time span of 1992–2000. Images had been classified to extract built-up, water, vegetation, and open land. Then they had adopted similar analytical approach as Sudhira et al. (2004), for characterizing the sprawl. This chapter also has similar limitations as in the preceding one (i.e., Sudhira et al. 2004).

Kumar et al. (2007) considered Indore city for a similar analysis. They analyzed the city using three temporal satellite remote sensing data for a period of 1990–2000. The growth patterns of built-up had been studied initially by dividing the area into four zones. The observations had been made with respect to each zone. Then, the study area had been divided into concentric circles of 1 km buffers; the growth patterns had been studied based on urban built-up density with respect to each circular buffer in all four zones. These observations had been integrated with road network to check the influence of infrastructure on haphazard urban growth. They also used entropy and population versus built-up growth for the identification of sprawl. This study had shown a novelty in dividing the zones and integrating the zones with road network.

A paper by Jat et al. (2008) is another example which analyzed the Ajmer city (Rajasthan) for the time span of 1977–2005 using eight temporal satellite remote sensing data. They classified the images using supervised maximum likelihood classifier to extract ten information classes. Then landscape metrics (such as, entropy, patchiness or landscape diversity, built-up density, population vs. built-up index) had been calculated. Landscape metrics had been extracted using a 5 × 5 moving window to characterize the sprawl. Finally, they modeled the urban growth for 2011 and 2051. This research was comprehensive in all respects.

A research by Jha et al. (2008) was aimed to study the spatial extent of urbanization in Haridwar, and patterns of periodic changes in urban development (systematic/random) in order to develop future plans. Remote sensing images had been used to map the spatial extent of urbanization for the years 1989, 1998, 2000, and 2002. Entropy was used to study the patterns of urban development (systematic or random) for the time span 1989–2002. The distributed entropy and relative mean entropy values had been evaluated considering two location factors: (i) urban development at peripheries of 1,000 m each from the center of the city, (ii) urban development at peripheries of 1,000 m each from the highway along the upper *Ganga* canal. This research, although not solely focused on sprawl, has shown novelty in using the entropy.

Another recent work, conducted by Farooq and Ahmad (2008), was aimed to analyze the urban sprawl around Aligarh city by using remote sensing data for a time span of 1971–2006. They used Survey of India topographical maps for 1971 and remote sensing images for the other three years (1989, 1999, and 2006). Then built-up areas had been mapped by means of visual inspection. Built-up growth and population data had been used to identify the sprawled area. In this chapter, rapid built-up growth and visual inspection of pattern remained the scales to identify the sprawl. The novelty of this chapter was in discussing the consequences of urban growth for the study area.

2.10 Similar Studies on Kolkata

Although the city of Kolkata is a very old and important urban area in India, analysis of urban growth and sprawl for this city was not found in the existing literature when this research was initiated. Kolkata urban agglomeration is the largest urban agglomeration in eastern India, and second largest in India with a population of 13.2 million as per 2001 census. This area contains more than 50 % of the total population of West Bengal. Therefore, while initiating this research, the city of Kolkata has been considered as a study area to understand the pattern and process of urban growth and sprawl.

A research by Richardson et al. (2000) found that cities in developing countries are becoming significantly more compact in spite of decelerating population growth and the beginnings of decentralization. Acioly and Davidson (1996) also reported that there was evidence that a general process of change was leading to more compact cities in developing countries. Whether the city of Kolkata supports their findings or contradicts them would be interesting to know through the current research.

2.11 Scope of the Research

The scope of this research, in view of the observations documented in this chapter, can be listed as:

(1) Analysis of urban growth and sprawl (from remote sensing data) for the city of Kolkata for the time span 1980–2010.
(2) Application of several existing models for the analysis of urban growth and sprawl for the city of Kolkata to evaluate their results and suitability.
(3) Modification of existing models (if require) to analyze the urban growth and sprawl for the city of Kolkata.
(4) Proposal of some simple (in contrast of complex) new approaches that require minimal set of input data to analyze the urban growth and sprawl. The analysis of urban growth is now essential for many layers of the administration, environmentalists, planners, stakeholders, and even individuals. It is not necessary that they will be an expert in remote sensing and GIS. Therefore,

simple approaches are preferred for quick and wide adaptation. Developing countries generally lack consistent and reliable data, especially spatial data (e.g., multi-temporal road network, land-use/land-cover maps, and parcel data). Census data are also highly generalized for the past. Therefore, analysis should not be based on a wide variety of data that are not possible to obtain easily, especially for public consumption.

(5) Identification of simple models that can characterize the sprawl in black-and-white. Relative measures, most often, fail to draw conclusion on sprawl and cannot be used in other areas reliably. These measures may serve the scientific purposes well, but, never can become a technology, because to interpret the results one has to be a scientist.

(6) Introduction of new zoning concept that is not dependent on the administrative block or buffer or circular/pie sectional divisions; rather is based on natural multi-temporal boundary of the city.

References

Acioly CC, Davidson F (1996) Density in urban development. Building Issues 8(3):3–25

Alberti M, Waddell P (2000) An integrated urban development and ecological simulation model. Integr Assess 1:215–227

Almeida CM, Monteiro AMV, Mara G, Soares-Filho BS, Cerqueira GC, Pennachin CSL, Batty M (2005) GIS and remote sensing as tools for the simulation of urban land-use change. Int J Remote Sens 26(4):759–774

Angel S, Sheppard SC, Civco DL (2005) The dynamics of global urban expansion. Transport and Urban Development Department, The World Bank, Washington, DC, p 200. http://www.citiesalliance.org/doc/resources/upgrading/urbanexpansion/worldbankreportsept2005.pdf

Angel S, Parent J, Civco D (2007) Urban sprawl metrics: an analysis of global urban expansion using GIS. Proceedings of ASPRS 2007 annual conference, Tampa, 7–11 May. http://clear.uconn.edu/publications/research/tech_papers/Angel_et_al_ASPRS2007.pdf

Armhein C (1995) Searching for the elusive aggregation effect: evidence from statistical simulations. Environ Plan A 27(1):105–119

Audirac I, Shermyen AH, Smith MT (1990) Ideal urban form and visions of the good life: Florida's growth management dilemma. J Am Plan Assoc 56(4):470–482

Axtell RL, Epstein JM (1994) Agent-based modeling: understanding our creations. *The Bulletin of the Santa Fe Institute*. Winter 28–32

Bae CHC, Richardson HW (1994) Automobiles, the environment and metropolitan spatial structure. Lincoln Institute of Land Policy, Cambridge

Banister D, Watson S, Wood C (1997) Sustainable cities: transport, energy, and urban form. Environ Plan B 24:125–143

Barnes KB, Morgan JM III, Roberge MC, Lowe S (2001) Sprawl development: its patterns, consequences, and measurement. A white paper, Towson University. http://chesapeake.towson.edu/landscape/urbansprawl/download/Sprawl_white_paper.pdf

Batty M (1976) Entropy in spatial aggregation. Geogr Anal 8:1–21

Batty M, Howes D (2001) Predicting temporal patterns in urban development from remote imagery. In: Donnay JP, Barnsley MJ, Longley PA (eds) Remote sensing and urban analysis. Taylor and Francis, London, pp 185–204

Benfield FK, Raimi M, Chen D (1999) Once there were greenfields: how urban sprawl is undermining America's environment. Economy and Social Fabric, The Natural Resources Defense Council, Washington, DC

Bhatta B (2010) Analysis of urban growth and sprawl from remote sensing data. Springer, Heidelberg, p 172
Bhatta B, Saraswati S, Bandyopadhyay D (2010a) Urban sprawl measurement from remote sensing data. Appl Geogr 30(4):731–740
Bhatta B, Saraswati S, Bandyopadhyay D (2010b) Quantifying the degree-of-freedom, degree-of-sprawl, and degree-of-goodness of urban growth from remote sensing data. Appl Geogr 30(1):96–111
Burchell RW, Shad NA, Lisotkin D, Phillips H, Downs A, Seskin S et al (1998) The costs of sprawl revisited. National Academy Press, Washington, DC
Calthorpe P, Fulton W, Fishman R (2001) The regional city: planning for the end of sprawl. Island Press, Washington, DC
Clapham WB Jr (2003) Continuum-based classification of remotely sensed imagery to describe urban sprawl on the watershed scale. Remote Sens Environ 86:322–340
Donnay JP, Barnsley MJ, Longley PA (2001) Remote sensing and urban analysis. In: Donnay JP, Barnsley MJ, Longley PA (eds) Remote sensing and urban analysis. Taylor & Francis, London, pp 3–18
Du Y, Guindon B, Cihlar J (2002) Haze detection and removal in high resolution satellite images with wavelet analysis. IEEE Trans Geosci Remote Sens 40:210–217
Duany A, Plater-Zyberk E, Speck J (2001) Suburban nation: the rise of sprawl and the decline of the American dream. North Point Press, New York
EEA (European Environment Agency) (2002) Towards an urban atlas: assessment of spatial data on 25 European cities and urban areas. EEA, Copenhagen
El Nasser H, Overberg P (2001) A comprehensive look at sprawl in America. McLean, Virginia. USA Today, February 22
Ewing R (1994) Characteristics, causes, and effects of sprawl: a literature review. Environ Urban Stud 21(2):1–15
Ewing R (1997) Is Los Angeles-style sprawl desirable? J Am Plan Assoc 63(1):107–126
Ewing R, Pendall R, Chen DDT (2002) Measuring sprawl and its impact. Smart Growth America, Washington, DC
Farley R, Frey WH (1994) Changes in the segregation of whites from blacks in the 1980 s: small steps toward a more integrated society. Am Sociol Rev 59:23–45
Farooq S, Ahmad S (2008) Urban sprawl development around Aligarh city: a study aided by satellite remote sensing and GIS. J Indian Soc Remote Sens 36:77–88
Fulton W, Pendall R, Nguyen M, Harrison A (2001) Who sprawls the most? how growth patterns differ across the US. The Brookings Institution Center on Urban and Metropolitan Policy, Washington, DC
Galster G (1991) Black suburbanization: has it changed the relative location of the races? Urban Aff Q 26:621–628
Galster G, Hanson R, Wolman H et al (2000) Wrestling sprawl to the ground: defining and measuring an elusive concept. Working paper of Fannie Mae, pp 1–38
Galster G, Hanson R, Wolman H, Coleman S, Freihage J (2001) Wrestling sprawl to the ground: defining and measuring an elusive concept. Hous Policy Debate 12(4):681–717
Geoghegan J, Wainger LA, Bockstael NE (1997) Spatial landscape indices in a hedonic framework: an ecological economics analysis using GIS. Ecol Econ 23(3):251–264
Geymen A, Baz I (2008) Monitoring urban growth and detecting land-cover changes on the Istanbul metropolitan area. Environ Monit Assess 136:449–459
Gordon P, Richardson HW (1997a) Are compact cities a desirable planning goal? J Am Plan Assoc 63(1):95–106
Gordon P, Richardson HW (1997b) Where's the sprawl? J Am Plan Assoc 63(2):275–278
Gustafson EJ (1998) Quantifying landscape spatial pattern: what is the state of the art? Ecosystems 1:143–156

Hardin PJ, Jackson MW, Otterstrom SM (2007) Mapping, measuring, and modeling urban growth. In: Jensen RR, Gatrell JD, McLean D (eds) Geo-spatial technologies in urban environments, 2nd edn. Springer, Berlin, pp 141–176

Hargis CD, Bissonette JA, David JL (1998) The behavior of landscape metrics commonly used in the study of habitat fragmentation. Landsc Ecol 13:167–186

Harvey RO, Clark WAV (1965) The nature and economics of urban sprawl. Land Econ 41(1):1–9

Hasse J (2004) A geospatial approach to measuring new development tracts for characteristics of sprawl. Landsc J 23(1):52–67

Hasse J, Lathrop RG (2003a) A housing-unit-level approach to characterizing residential sprawl. Photogramm Eng Remote Sens 69(9):1021–1030

Hasse JE, Lathrop RG (2003b) Land resource impact indicators of urban sprawl. Appl Geogr 23:159–175

Hedblom M, Soderstrom B (2008) Woodlands across Swedish urban gradients: status, structure and management implications. Landsc Urban Plan 84:62–73

Herold M, Goldstein N, Clarke KC (2003) The spatio-temporal form of urban growth: measurement, analysis and modeling. Remote Sens Environ 86(3):286–302

Herold M, Roberts D, Gardner M, Dennison P (2004) Spectrometry for urban area remote sensing—development and analysis of a spectral library from 350 to 2400 nm. Remote Sens of Environ 91:304–319

Herold M, Couclelis H, Clarke KC (2005a) The role of spatial metrics in the analysis and modeling of urban change. Comput Environ Urban Syst 29(4):369–399

Herold M, Hemphill J, Dietzel C, Clarke KC (2005b) Remote sensing derived mapping to support urban growth theory. Proceedings of ISPRS, vol XXXVI-8/W27. http://www.isprs.org/proceedings/XXXVI/8-W27/herold_hemphill_etal.pdf

HUD (1999) The state of the cities. US Department of Housing and Urban Development, Washington, DC

Jat MK, Garg PK, Khare D (2008) Modeling urban growth using spatial analysis techniques: a case study of Ajmer city (India). Int J Remote Sens 29(2):543–567

Jensen JR (2005) Introductory digital image processing: a remote sensing perspective. Prentice-Hall, Upper Saddle River

Jensen JR, Cowen DC (1999) Remote sensing of urban/suburban infrastructure and socioeconomic attributes. Photogramm Eng Remote Sens 65(5):611–622

Jha R, Singh VP, Vatsa V (2008) Analysis of urban development of Haridwar, India, using entropy approach. KSCE J Civil Eng 12(4):281–288

Jiang F, Liu S, Yuan H, Zhang Q (2007) Measuring urban sprawl in Beijing with geo-spatial indices. J Geog Sci. doi:10.1007/s11442-007-0469-z

Johnson MP (2001) Environmental impacts of urban sprawl: a survey of the literature and proposed research agenda. Environ Plan A 33:717–735

Kline JD (2000) Comparing states with and without growth management analysis based on indicators with policy implications comment. Land Use Policy 17:349–355

Kumar JAV, Pathan SK, Bhanderi RJ (2007) Spatio-temporal analysis for monitoring urban growth—a case study of Indore city. J Indian Soc Remote Sens 35(1):11–20

Kundu A, Basu S (1999) Words and concepts in urban development and planning in India: an analysis in the context of regional variation and changing policy perspectives. MOST City Words Project, Working paper no 4, UNESCO. www.unesco.org

Lang RE (2003) Open spaces, bounded places: does the American West's arid landscape yield dense metropolitan growth? Hous Policy Debate 13(4):755–778

Lata KM, Rao CHS, Prasad VK, Badarianth KVS, Rahgavasamy V (2001) Measuring urban sprawl: a case study of Hyderabad. GISdevelopment 5(12):26–29

Ledermann RC (1967) The city as a place to live. In: Gottmann J, Harpe RA (eds) Metropolis on the move: geographers look at urban sprawl. Wiley, New York

Lessinger J (1962) The cause for scatteration: some reflections on the National Capitol Region plan for the year 2000. J Am Inst Plan 28(3):159–170
Li X, Yeh AGO (2004) Analyzing spatial restructuring of landuse patterns in a fast growing region remote sensing and GIS. Landsc Urban Plan 69:335–354
Longley PA, Mesev V (2000) On the measurement and generalization of urban form. Environ Plan A 32:473–488
Longley PA, Barnsley MJ, Donnay JP (2001) Remote sensing and urban analysis: a research agenda. In: Donnay JP, Barnsley MJ, Longley PA (eds) Remote sensing and urban analysis. Taylor and Francis, London, pp 245–258
Malpezzi S (1999) Estimates of the measurement and determinants of urban sprawl in U.S. Metropolitan areas. Center for Urban Land Economics Research, University of Wisconsin, Madison
Martinuzzi S, Gould WA, Gonzalez OMR (2007) Land development, land use, and urban sprawl in Puerto Rico integrating remote sensing and population census data. Landsc Urban Plan 79:288–297
Masek JG, Lindsay FE, Goward SN (2000) Dynamics of urban growth in the Washington DC metropolitan area, 1973–1996, from Landsat observations. Int J Remote Sens 21:3473–3486
McGarigal K, Marks BJ (1995) FRAGSTATS: spatial pattern analysis program for quantifying landscape structure. USDA Forest Service General Technical Report PNW-351
McGarigal K, Cushman SA, Neel MC, Ene E (2002) FRAGSTATS: spatial pattern analysis program for categorical maps. www.umass.edu/landeco/research/fragstats/fragstats.html
NASA (2001) Satellite maps provide better urban sprawl insight. NASA News Release, 2 June 2001
Nelson AC (1999) Comparing states with and without growth management: analysis based on indicators with policy implications. Land Use Policy 16:121–127
O'Neill RV, Krummel JR, Gardner RH, Sugihara G, Jackson B, Deangelis DL, Milne BT, Turner MG, Zygmunt B, Christensen SW, Dale VH, Graham RL (1988) Indices of landscape pattern. Landsc Ecol 1:153–162
Openshaw S (1984) The modifiable areal unit problem. Geo Books, Norwich
Openshaw S (1991) A spatial analysis research agenda. In: Masser I, Blakemore M (eds) Handling geographical information: methodology and potential applications. Wiley, New York, pp 18–37
Openshaw S, Alvandies S (1999) Applying geocomputation to the analysis of spatial distributions. In: Longley P, Goodchild M, Maguire D, Rhind D (eds) Geograhpic information systems: principles and technical issues, vol 1, 2nd edn. Wiley, New York
OTA (1995) The technological reshaping of metropolitan America. DCUS Congress Office of Technology Assessment, Washington
Páez A, Scott DM (2004) Spatial statistics for urban analysis: a review of techniques with examples. GeoJournal 61(1):53–67
Paolini L, Grings F, Sobrino J, Jiménez Muñoz J, Karszenbaum H (2006) Radiometric correction effects in Landsat multi-date/multi-sensor change detection studies. Int J Remote Sens 27:685–704
Parker DC, Evans TP, Meretsky V (2001) Measuring emergent properties of agent-based landuse/landcover models using spatial metrics. Proceedings of 7th annual conference of the international society for computational economics. http://php.indiana.edu/~dawparke/parker.pdf
Parker DC, Manson SM, Janssen MA, Hoffmann MJ, Deadman P (2003) Multi-agent systems for the simulation of land-use and land-cover change: a review. *Annu Assoc Amer Geograph* 93(2):314–337
Peiser R (1989) Density and urban sprawl. Land Econ 65(3):193–204
Pendall R (1999) Do land-use controls cause sprawl? Environ Plan B 26(4):555–571
Prenzel B (2004) Remote sensing-based quantification of land-cover and land-use change for planning. Prog Plan 61:281–299

RERC (Real Estate Research Corporation) (1974) The costs of sprawl: environmental and economic costs of alternative residential patterns at the Urban Fringe. US Government Printing Office, Washington, DC
Richardson HW, Bae CC, Baxamusa MH (2000) Compact cities in developing countries: assessment and implications. In: Mike J, Burgess R (eds) Compact cities: sustainable urban forms for developing countries. Spon Press, London, pp 1–25
Riitters KH, O'Neill RV, Hunsaker CT, Wickham JD, Yankee DH, Timmins SP, Jones KB, Jackson BL (1995) A factor analysis of landscape pattern and structure metrics. Landsc Ecol 10(1):23–39
Roca J, Burnsa MC, Carreras JM (2004) Monitoring urban sprawl around Barcelona's metropolitan area with the aid of satellite imagery. Proceedings of geo-imagery bridging continents, XXth ISPRS congress, 12–23 July, Istanbul, Turkey. http://www.isprs.org/publications/related/semana_geomatica05/front/abstracts/Dimecres9/T07.pdf
Sierra Club (1998) Sprawl: the dark side of the American dream. Research report. http://www.sierraclub.org/sprawl/report98/report.asp
Singh A (1989) Digital change detection techniques using remotely sensed data. Int J Remote Sens 10(6):989–1003
Smith DM (1975) Patterns in human geography. David & Charles, Newton Abbot, p 373
Sokhi BS, Sharma ND, Uttarwar PS (1989) Satellite remote sensing in urban sprawl mapping and monitoring: a case study of Delhi. J Indian Soc Remote Sens 17(3):57–69
Sudhira HS, Ramachandra TV (2007) Characterising urban sprawl from remote sensing data and using landscape metrics. Proceedings of 10th international conference on computers in urban planning and urban management, Iguassu Falls, 11–13 July. http://eprints.iisc.ernet.in/11834/
Sudhira HS, Ramachandra TV, Raj KS, Jagadish KS (2003) Urban growth analysis using spatial and temporal data. J Indian Soc Remote Sens 31(4):299–311
Sudhira HS, Ramachandra TV, Jagdish KS (2004) Urban sprawl: metrics, dynamics and modelling using GIS. Int J Appl Earth Obs Geoinf 5:29–39
Sutton PC (2003) A scale-adjusted measure of "urban sprawl" using nighttime satellite imagery. Remote Sens Environ 86:353–369
Taragi RCS, Pundir PPS (1997) Use of satellite data in urban sprawl and land use studies—a case of Lucknow city. J Indian Soc Remote Sens 25(2):113–118
Thakur B, Parai A (1993) A review of recent urban geographic studies in India. *GeoJournal* 29(2):187–196
The Brookings Institution (2002) The growth in the Heartland: challenges and opportunities for Missouri. The Brookings Institution Center on Urban and Metropolitan Policy, Washington, DC
Thomas RW (1981) Information statistics in geography, geo abstracts. University of East Anglia, Norwich, p 42
Torrens PM (2000) How cellular models of urban systems work (1. Theory). Working paper series 28. Centre for Advanced Spatial Analysis, London. http://eprints.ucl.ac.uk/1371/
Torrens PM (2008) A toolkit for measuring sprawl. Appl Spatial Anal 1:5–36
Tsai Y (2005) Quantifying urban form: compactness versus 'sprawl'. Urban Stud 42(1):141–161
Turner MG, Gardner RH, O'Neill RV (2001) Landscape ecology in theory and practice: pattern and process. Springer, New York, p 401
USA Today (2001) A comprehensive look at sprawl in America. http://www.usatoday.com/news/sprawl/main.htm
USEPA (United States Environmental Protection Agency) (2001) Our built and natural environments. A technical review of the interactions between land use, transportation, and environmental quality. US Environmental Protection Agency. http://www.smartgrowth.org/pdf/built.pdf
Uttarwar PS, Sokhi BS (1989) Remote sensing application in urban fringe study: a case study–Delhi. J Indian Soc Remote Sens 17(3):43–56

Wang W, Zhu L, Wang R, Shi Y (2003) Analysis on the spatial distribution variation characteristic of urban heat environmental quality and its mechanism—a case study of Hangzhou city. Chin Geogr Sci 13(1):39–47
Wilson EH, Hurd JD, Civco DL, Prisloe S, Arnold C (2003) Development of a geospatial model to quantify, describe and map urban growth. Remote Sens Environ 86(3):275–285
Yeh AGO, Li X (2001) Measurement and monitoring of urban sprawl in a rapidly growing region using entropy. Photogramm Eng Remote Sens 67(1):83–90
Zhu M, Xu J, Jiang N, Li J, Fan Y (2006) Impacts of road corridors on urban landscape pattern: a gradient analysis with changing grain size in Shanghai, China. Landsc Ecol 21:723–734

Chapter 3
Methodology

Abstract This chapter describes the data that have been used to perform the research, and the methodology adopted for the analysis of urban growth and sprawl. Methodology starts from image registration, then describes map encoding, image classification, accuracy assessment, determination of natural boundary, subdivision of natural boundary, calculation of built-up area, and finally analyses. The main problem faced in this research was lack of temporal ancillary data sets that were essential for many of the widely practiced metrics. Therefore, the intention was to use some metrics that are less demanding in terms of data and computation. However, it has been found that most of these metrics are inferior in capturing insights of urban growth and sprawl. In this chapter, therefore, most of these metrics have been modified, either to overcome their limitations or to fit the study area. This chapter is also aimed to justify the methods adopted in the current research. However, before proceeding through this chapter, one may refer to Appendix A for an overview of the study area.

Keywords Urban growth · Sprawl · Remote sensing mapping · GIS · Encoding hardcopy map · Encoding vector map · Remotely sensed image classification · City extension · Households and urban growth

3.1 Data

The following multi-temporal remote sensing imageries have been used to extract built-up areas of Kolkata:

- Landsat Multispectral Scanner (MSS) image (path 148, rows 44 and 45) from 16 January 1980;
- Landsat Thematic Mapper (TM) image (path 138, row 44) from 14 November 1990;

B. Bhatta, *Urban Growth Analysis and Remote Sensing*, SpringerBriefs in Geography, 33
DOI: 10.1007/978-94-007-4698-5_3, © The Author(s) 2012

Table 3.1 Spectral details of the satellite imageries

Landsat MSS		Landsat TM and Landsat ETM +		Resourcesat-1 LISS-III	
Bands	Spectral resolution (μm)	Bands	Spectral resolution (μm)	Bands	Spectral resolution (μm)
4	0.5–0.6	1	0.45–0.52		
5	0.6–0.7	2	0.52–0.60	2	0.52–0.59
6	0.7–0.8	3	0.63–0.69	3	0.62–0.68
7	0.8–1.1	4	0.76–0.90	4	0.77–0.86
		5	1.55–1.75	5	1.55–1.70
		7	2.08–2.35		

- Landsat Enhanced Thematic Mapper Plus (ETM+) image (path 138, row 44) from 17 November 2000;
- Indian Remote Sensing (IRS) Resourcesat-1 Linear Imaging Self-scanning Sensor (LISS) III image (path 108, row 56) from 03 February 2010.

Landsat images were obtained from the National Aeronautics and Space Administration, United States of America. The LISS-III image has been procured from the National Remote Sensing Centre, India. Spectral details of the aforementioned imageries are furnished in Table 3.1. Thermal bands of Landsat sensors have not been considered for the analysis owing to their coarser spatial resolutions (120 m for TM and ETM+, and 240 m for MSS) compared to the optical bands. Furthermore, unlike the optical bands that give us the measures of percent reflectance, thermal bands provide information on relative radiant temperature. Mixing of these two properties may cause spectral confusion.

The spatial resolutions of the imageries are as follows: Landsat MSS has a nominal spatial resolution of 79 × 57 m, Landsat TM and ETM+ have spatial resolution of 30 × 30 m, and LISS III has a spatial resolution of 23.5 × 23.5 m. Landsat MSS image has been obtained as a 60 × 60 m resampled image.

The hardcopy of the Kolkata Municipal Corporation (KMC) *Ward Map* has been used to generate the vector maps. In addition to the hardcopy ward map of KMC, one panfused QuickBird imagery of KMC area (from the year 2005, having spatial resolution of 0.6 × 0.6 m) has also been referred for accurate mapping of KMC ward boundaries. This image has been referred from Google Earth Pro (http://earth.google.com).

A total of 10 ground control points (GCP), surveyed with Global Positioning System (GPS) receiver, have been used to make the satellite imageries geographically referenced.

Census data (*number of households* and *populations*) for the years 1981, 1991, and 2001 have also been referred for analytical purposes (Census 1981, 1991, 2001) (listed in Appendix B).

3.2 Software Used

The following software have been selected to execute the research:

- *ERDAS Imagine* for digital processing of remotely sensed data;
- *ArcGIS* as a GIS software;
- *Microsoft Excel* for statistical analysis.

ERDAS and ArcGIS software have been chosen because they support the file formats of one another. However, any standard remote sensing image processing software and GIS software can be selected to execute such type of study instead of ERDAS and ArcGIS.

3.3 Image Registration

Image registration or image-to-image geocorrection fits the coordinate system of one image to that of a second image of the same geographic area. Image registration involves matching the coordinate systems of two digital images with one image acting as a reference image and the other as the image to be rectified (Bhatta 2008). It involves rearrangement of the input pixels onto a new grid. Polynomial transformation equations are used to convert the source coordinates into rectified coordinates (ERDAS 2008).

Satellite imageries have been obtained as the standard product, which means, geometrically and radiometrically corrected. However, owing to several standards and references used by the image supplying agencies, the overlay of the images generally does not match with considerable accuracy. To solve this problem, the panfused QuickBird image has been geographically referenced with evenly distributed 10 GCPs within the study area. GCPs have been collected using the GPS receiver (manufacturer: *Trimble*, model: *GeoXH 6000*) with an accuracy of 1.5 m. Then, other satellite images have been co-registered so that the overlay matches with subpixel accuracy (maximum root mean square error $\approx$0.41). Nearest-neighbor resampling has been used to transform the imageries so that the original pixel value is retained. First-order polynomial transformation equations (ERDAS 2008) have been used to georeference/register the imageries.

Images from different sensors have differences in their spatial resolutions. One approach to encounter this problem is resampling the higher resolution images to match the resolution of the lower resolution image. But resampling of an image to change the pixel size either averages the neighboring pixel values (in case of bi-linear or bi-cubic) or pixel dropout/duplication occurs (in case of nearest neighbor); in addition, spatial detail also decreases (ERDAS 2008). Averaging of neighboring pixel values produces mixed pixels, which sacrifices the details to be extracted via classification. Pixel dropout/duplication also sacrifices the details or introduces errors. Many of the researchers prefer to resample the images to make

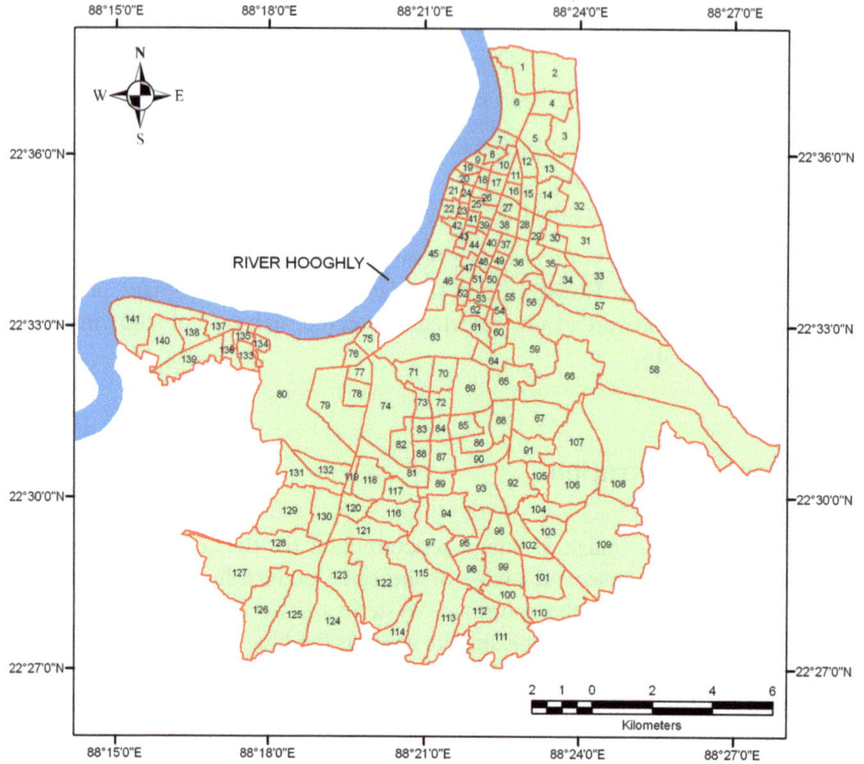

Fig. 3.1 Ward map of KMC as on 23 July 2010 (numerals show the ward numbers) (Courtesy: KMC)

them of same size in all images. This is especially preferred when one wants to perform mathematical operations among the images. However, the cases where we are not performing any mathematical operations among the images, such as the current research, we should not resample the images. In this research we need the built-up statistics that can be calculated independently. Therefore, the images have been kept without changing the pixel size despite the varying accuracy level of classification with the different spatial, spectral, and radiometric resolutions.

3.4 Encoding Hardcopy Map

The hardcopy ward map of KMC, as shown in Fig. 3.1, has been scanned with drum scanner to convert in digital raster image. This image has been registered with the satellite imageries using the same approach as mentioned in the Sect. 3.3. There were some mismatches (100–200 m) among the administrative boundaries in the considered images (scanned image and satellite imageries). This problem could not be solved because the errors were at the source of the scanned image.

Table 3.2 Initial feature classes and attributes for encoding vector data

Feature class	Attribute fields
KMC borough boundaries	ID
	Borough No.
	Number of households in 1981
	Number of households in 1991
	Number of households in 2001
	Population in 1981
	Population in 1991
	Population in 2001
River hooghly	ID
	Name

3.5 Encoding Vector Maps

A *personal geodatabase* has been created in the ArcGIS software. *Geodatabase* is a relational database that stores geographic data as well as attribute data. It is an object-oriented data model introduced by Environmental Systems Research Institute Inc., and used to store spatial and attribute data and the relationships that exist among them. A *personal geodatabase* has the *.mdb* file extension (similar to a format used by Microsoft Access).

Within the aforementioned geodatabase, one *feature dataset* has been created. Feature data set is a collection of feature classes with user-defined spatial relationships and topologies; it is stored within a geodatabase. In ArcGIS, a collection of *feature classes* are stored together that share the same spatial reference; that is, they share a coordinate system, and their features fall within a common geographic area. Feature classes with different geometry types and/or classes may be stored in a feature data set. A *feature class* is a collection of geographic features that share the same geometry type (such as point, line, or polygon), and the same attribute fields for a common area.

Feature classes, initially, have been created for KMC boroughs and for the river Hooghly. Once the spatial data are entered for the boroughs, the borough boundaries can be merged to generate the KMC boundary. Attribute data for the boroughs can also be summarized to generate the attributes of KMC. Initial feature classes and their associated attribute fields are listed in Table 3.2.

In the next step, the encoded scanned map has been used to generate the vector borough map of KMC along with the river Hooghly as shown in Fig. 3.2. Both the scanned ward map of KMC and the panfused QuickBird image have been referred to generate the vector map of KMC boroughs. River Hooghly has been drawn with reference to the LISS-III satellite imagery of year 2010. Attribute data for the fields listed in Table 3.2 have been entered within the geodatabase. Appendix B (Table 3.2) lists the details of census data that have been entered for the boroughs.

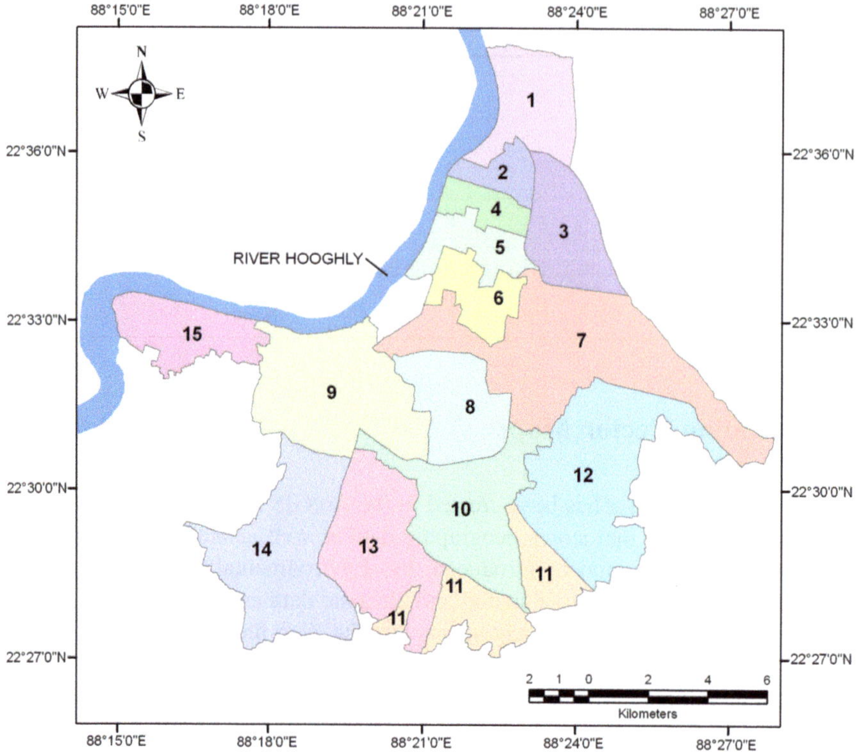

Fig. 3.2 Borough map of KMC as on 23 July 2010 (numerals show the borough numbers)

3.6 Image Classification

Remotely sensed images can be digitally classified via *supervised* or *unsupervised* approaches. Supervised classification is much more accurate for mapping information classes, but largely depends on the cognition and skills of the image analyst (Bhatta 2008); whereas unsupervised classification is more computer-automated. Supervised classification approach is the best way for extracting a single class from the imagery (Jensen 2005). Since the current study extracts a single class (built-up), supervised is the preferred approach. It is worth mentioning that impervious surfaces (such as built-up) are promoted as useful environmental indicators (Arnold and Gibbons 1996; Barnes et al. 2000; Schueler 1994), and one environmental condition that impervious surfaces clearly indicate is urbanization (Barnes et al. 2001). If the study is concerned only with urban growth (only the classes of urban and non-urban use are under consideration), a simple binary classification of remotely sensed data is enough (Torrens and Alberti 2000; Barnes et al. 2001; Epstein et al. 2002).

There are a number of verities in decision rules used in supervised classification (Jensen 2005; ERDAS 2008), which may be divided into two main categories: *parametric* and *nonparametric*. A parametric decision rule is trained by the parametric signatures. These signatures are defined by the mean vector and covariance matrix for the data file values of the pixels in the signatures (ERDAS 2008). When a parametric decision rule is used, every pixel is assigned to a class, since the parametric decision space is continuous. A nonparametric decision rule is not based on statistics. If a pixel is located within the boundary of a nonparametric signature, then this decision rule assigns the pixel to the signature's class. Basically, a nonparametric decision rule determines whether or not the pixel is located inside of the nonparametric signature boundary (inside the specified spectral limit). Nonparametric techniques are sometimes termed 'robust', because they can be applied with a wide variety of class distribution, if class signatures are distinct to begin with (ERDAS 2008).

Co-registered images, therefore, have been classified by using nonparametric feature-space classifier (ERDAS 2008) to extract the built-up area along with other impervious surfaces (such as runways) to determine the built-up class. A fundamental difference between the feature space classifier and the other traditional methods is that it is based on nonparametric signature. The decisions made in the classification process have no dependency on the statistics of the pixels. This helps to improve classification accuracies for specific nonnormal classes, such as urban (ERDAS 2008).

3.7 Assessment of Classification Accuracy

Accuracy assessment of remote sensing product is a feedback system for checking and evaluating the objectives and the results. Accuracy assessments determine the correctness of the classified image that is based on pixel groupings. Accuracy is a measure of the agreement between a standard that is assumed to be correct and an image classification of unknown quality. If the image classification corresponds closely with the standard, it is said to be accurate. Accuracy assessment for remote sensing classification is commonly based on using an error matrix, or confusion table, which needs reference map or high resolution image, or 'ground truthing' data to support (Bhatta 2008). There are several different ways in which assessment of accuracy can be accomplished. One method is to compare the classified image to a reference image. A random set of points is generated and classification results are compared with the true information classes in the reference image. A second method to perform accuracy assessment involves using a GPS receiver. Again, a random set of points is generated over the classified image. Ground truthing would be performed by going into the field at the location of each randomly generated point. The classification results would, then, be compared to actual land cover at each point's location.

However, when undertaking change detection using multi-temporal images, it is often difficult to perform the accuracy assessment via 'traditional' method that typically requires simultaneous collection of reference data. When multi-temporal historical images are in consideration, it is often difficult to obtain corresponding reference maps or high resolution imageries, especially in developing countries. In case of Kolkata Metropolitan Area (KMA), no such reference maps (showing built-up) have been found for the years 1980, 1990, and 2000. Therefore, *rule-based rationality evaluation* (Liu and Zhou 2004) has been performed for the assessment of accuracy of the classified imagery.

In case of rule-based rationality evaluation, three rules have been applied to determine the rationality of the change trajectory for each sample. Let A denote the case that 'the pixel is correctly classified'. When A is rejected, it means 'the pixel is not correctly classified', thus it is the classification error. Land cover classes are represented as C_b = built-up class, and C_{nb} = nonbuilt-up class. The detected change trajectory between the land cover types is denoted as, for example, $T(C_{nb}, C_b)$ means 'change from non built-up to built-up', $T(C_b, C_{nb})$ means 'change from built-up to non built-up'. If we let n denote the number of detected categorical changes over the four monitoring dates from 1980 to 2010, we have:

$$n = 0 \quad or \quad 1 \tag{3.1}$$

where, the case of $n = 0$ refers to 'no change', while $n = 1$ refers to the fact that the category of the sample has changed for a specific detection period (e.g., 1980–1990).

Rule I: IF $n = 0$ THEN accept A

Rule II: IF $n = 1$ AND $T(C_{nb}, C_b)$ THEN accept A

Rule III: IF $n = 1$ AND $T(C_b, C_{nb})$ THEN reject A (built-up is irreversible to nonbuilt-up)

The meaning of *Rule I* is obvious. If the pixel is classified as the same land cover type for all monitoring dates, then there is no change and the pixel is regarded as correctly classified. The built-up area of KMA is increasing; it is happening because other land cover types are transforming to built-up areas and this process cannot be reversed. *Rule II*, therefore, states that if change is detected from non built-up to built-up, then the change is regarded as a 'true' case so that the pixel is correctly classified. *Rule III*, on the other hand, defines that if the reverse process was detected (i.e. change from built-up to nonbuilt-up), the change would be unlikely so that the pixel is not correctly classified. However, the imageries have different spatial resolutions. Owing to this problem and misregistration problem, imageries do not match exactly (Fig. 3.3). Therefore, the points should not be considered which fall in the confusion areas (areas of pixel mismatch).

The accuracy assessment has been performed for 500 random locations within the study area. Where 87 points have been found in the confusion areas (areas of pixel mismatch) and rejected without assessment. Among the remaining 413 locations 47 samples have been found as changed irreversibly (i.e. from built-up to

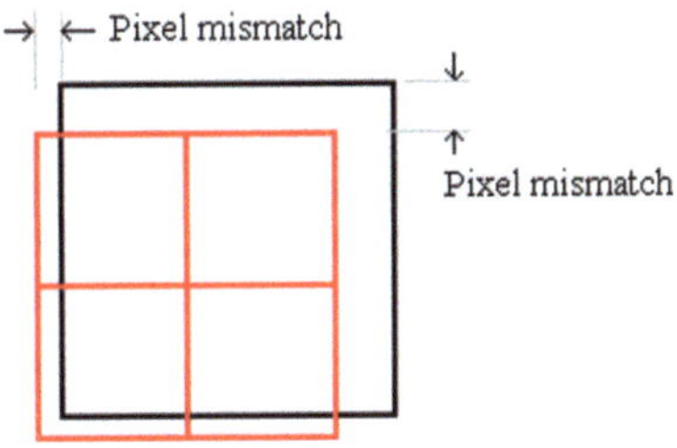

Fig. 3.3 Pixel mismatch due to different pixel sizes and misregistration

nonbuilt-up), constituting 11.38 % of the total. Thus, the result shows that the overall classification accuracy was at least 88.62 %.

An overall classification accuracy of 85 % is commonly considered sufficient for a remote sensing data product (Anderson et al. 1976). Therefore, it can be said that the accuracy assessment has resulted in sufficient confidence to proceed with the classified imageries. If the accuracy assessment does not provide satisfactory result, then the images are required to be reclassified with different signature sets or with different classification rules and the reassessment of accuracy should be done.

Thematic accuracy obviously directly influences the further analysis of the map with spatial metrics (Barnsley and Barr 2000). Although an overall classification accuracy of 85 % is commonly considered sufficient for a remote sensing data product (Anderson et al. 1976), post-classification comparison of two such temporal images will produce an accuracy around 72 % ($0.85 \times 0.85 \times 100$) that may not be acceptable in many of the instances where extremely high accuracies are needed. No guidance on minimum required level of accuracy by the local policy makers has been found in the existing literature which would be better to justify the acceptance of such accuracy level in this research.

3.8 Subsetting of Remote Sensing Data

Subsetting (or spatial subsetting) refers to breaking out a portion of a large image into one or more smaller pieces. Often, image files contain areas much larger than a particular study area. In such cases, it is helpful to reduce the extent of the image to include only the area of interest. This approach not only eliminates the extraneous data in the file, but also speeds up computer processing owing to the smaller amount of data to process.

Each of the scenes collected for this study contained a huge additional area outside of the KMC area. Therefore, co-registered images have been clipped to cut small pieces that represent KMC area. However, the unclipped data have also been kept for the analysis that considers the natural boundary of Kolkata rather than KMC.

3.9 Determining the Natural Boundary of the City

As mentioned in Sect. 2.6 (Chap. 2), analysis of urban growth can be performed in consideration of administrative as well as natural boundary of the city. Natural boundary is more preferred if the analysis does not require census or socioeconomic data. However, as mentioned in Sect. 2.7 (Chap. 2), determining the natural extent of a city is a difficult task, especially from remote sensing data.

In this research, in consideration of physical or morphological criteria, a simple approach has been adopted to determine the extent of the city. A circular neighborhood search from the urban core, for each date under analysis, has been conducted on the classified images to derive the contiguous urban area; and this boundary has been considered as the physical extent of the city (reported earlier in Bhatta (2009b)). The neighborhood search determines which pixels will be considered contiguous to the *seed* (selected) pixel. Any neighboring pixel that meets all selection criteria (in this case, belonging to built-up class) is accepted and thus, itself, becomes a seed pixel (ERDAS 2008). This has resulted in a contiguous area for each date under analysis. These contiguous areas can be considered as temporal boundaries or extents of the city, referred as to "city-extents" (Fig. 3.4). Discontinuous or scattered built-up pixels, outside of these city extents, have not been included since they are generally considered as isolated developments, not connected with the core urban area. However, nonurban islands within the contiguous built-up pixels have been included, because the intention was to demarcate the extent (boundary) of the city, not to find out the contiguous built-up pixels. Encouragement of urban growth within this extent will force *infill* growth that is generally considered as remedies to sprawl (Wilson et al. 2003). This approach can eliminate the problems associated with determining the urban boundary in a rural to urban gradient.

3.10 Subdivision of Natural City Extent

In case of analyzing the intra-city variations of urban growth and sprawl, it is necessary to subdivide the extent of the city. As discussed in Sect. 2.8 (Chap. 2), existing approaches to subdivide the natural city extent has several limitations. This research proposes a new approach to subdivide the natural extent of the city. It considers the multi-temporal natural extents of the city for the subdivision of study area.

The city extent of 1980 has been considered as *Zone 1*, the difference area between the city extents of 1980 and 1990 has been considered as *Zone 2*, difference between 1990 and 2000 has been considered as *Zone 3*, and the difference between 2000 and 2010 has been regarded as *Zone 4*. Figure 3.5 represents these zones spatially. Advantages of this zoning concept include: (1) It is not influenced by the spatial resolutions at a high degree, (2) it considers the natural growth of the city in zoning process rather than any artificial or hypothetical shapes, and (3) it does not include large rural (or other) areas within it. Sometimes, a large area may

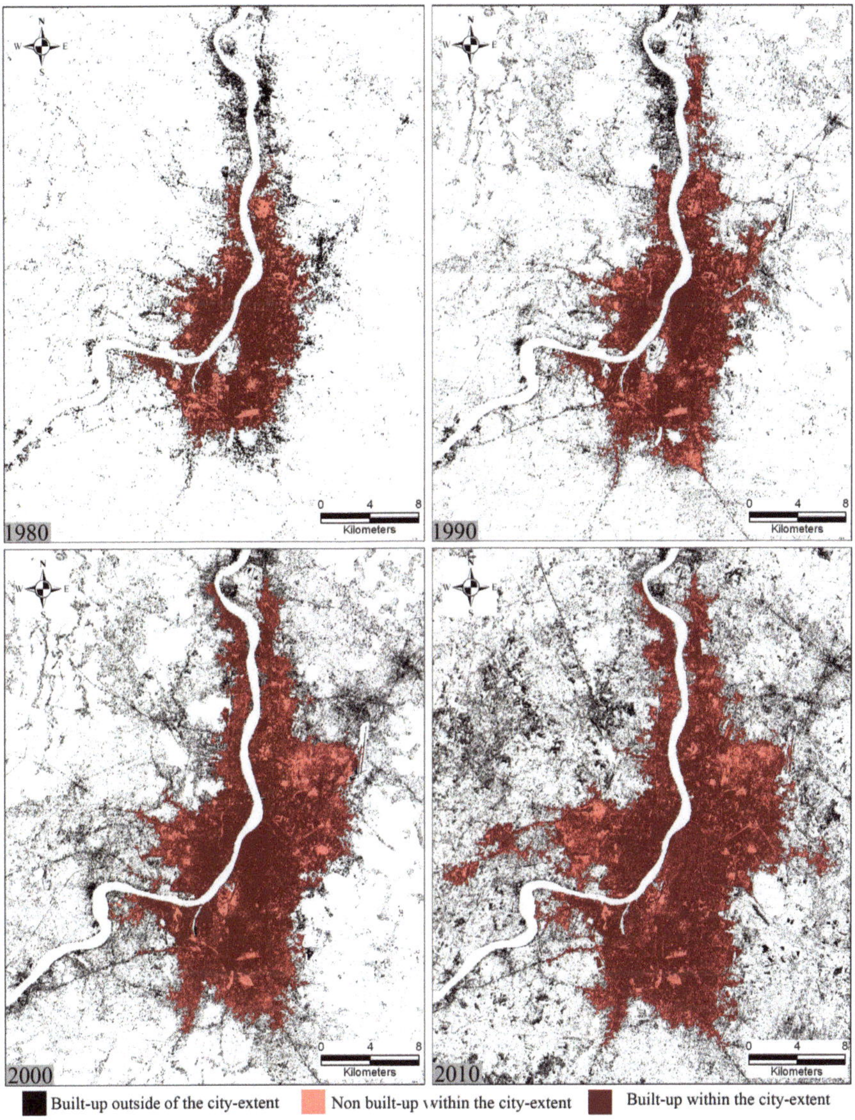

Fig. 3.4 Natural extents of the city for different dates

be left blank and the built-up may jump that area as because it is unsuitable for development or protected from development purposefully. Administrative, or circular/pie, or buffer subdivisions will consider these areas in the analysis and the area will artificially indicate a sprawl. The subdivision approach adopted in this research can avoid such type of large areas. Figure 3.5 clearly shows that River Hooghly and East-Kolkata wetlands have been excluded.

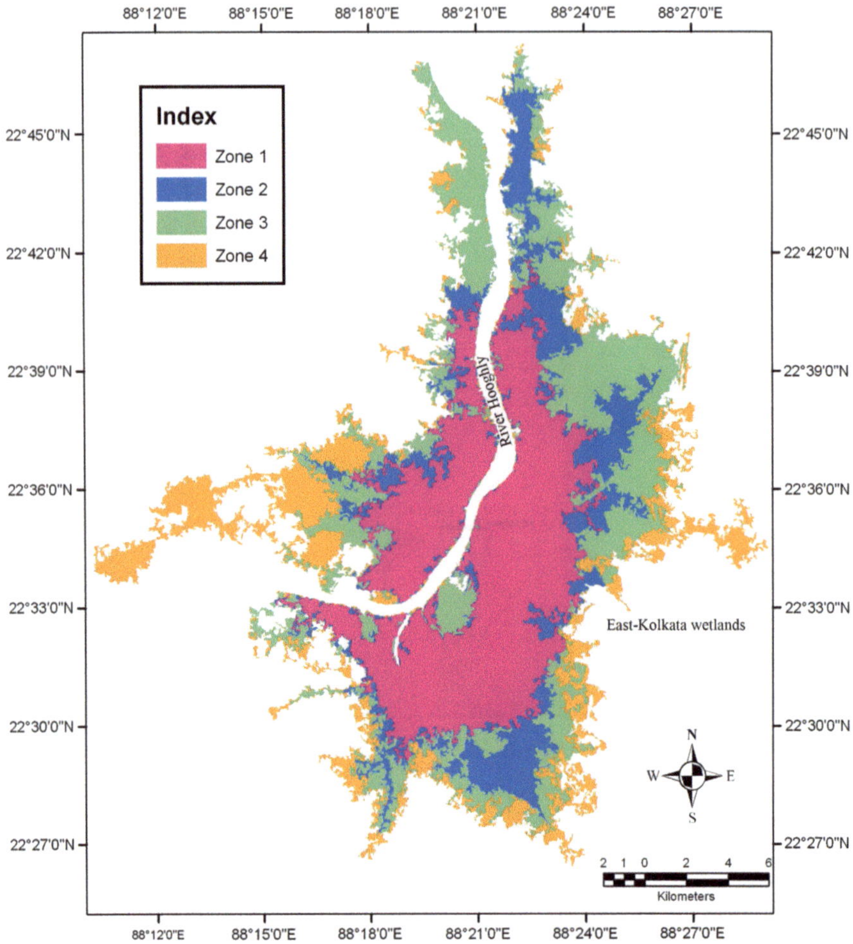

Fig. 3.5 Zoning concept of the study area in consideration of natural city extents

3.11 Calculating the Built-up Area

Built-up areas for each borough in the KMC and for each temporal image have been calculated by clipping the classified images with the respective vector, and then multiplying the number of pixels in each zone (borough) by the pixel size in ERDAS Imagine software (Appendix C).

In the same way, built-up areas in each natural extent of the city and in each subdivided zone have also been calculated (Appendix C). The results, then, have been used for the analysis of urban growth and sprawl.

3.12 Generation of KMC Boundary

Borough boundaries of KMC (along with attribute data such as census data and built-up data) have been summarized to generate the KMC boundary and its associated attribute data. This summarization has been performed in ArcGIS software. This process has resulted in a vector map of KMC and its associated attributes.

The methodology described up to this point can be represented in a simple flowchart as shown in Fig. 3.6.

3.13 Analysis Based on Administrative Boundary

In case of administrative boundary, the borough boundaries of KMC have been considered. Ward boundaries could not be considered in this analysis, because areas of some wards are too small in respect of the spatial resolutions of the imagery. For example, Ward 23 has an area of 16.59 ha, which can be covered by only 44 pixels of Landsat MSS image. This clearly shows that the spatial resolutions of the images under analysis are insignificant for such a detailed analysis. At least 4–12 m spatial resolution would be preferred at this scale (Neer 1999).

It would be better if the KMA could also be analyzed in consideration of municipal boundaries within this area. However, it was not possible because of several delimitations and inclusion of new municipalities within this area. Several outgrowths which have become towns recently, had not been accounted earlier for with any individual urban unit (Census 1981, 1991). Therefore, population figures of KMA municipalities do not tally by taking totaling the individual unit. Furthermore, no such readily available multi-temporal maps have been found that shows each administrative urban as well as rural unit within the KMA. However, the natural extent of the city automatically includes the entire KMA.

3.13.1 Built-up Area and Urban Growth

The percentage of an area covered by impervious surfaces such as asphalt and concrete is a straightforward measure of development (Barnes et al. 2001). It can safely be considered that developed areas have greater proportions of impervious surfaces (i.e. the built-up areas) compared to the lesser developed areas (Sudhira et al. 2004).

Therefore, the built-up areas for each borough as well as entire KMC and for each date under analysis have been calculated. The percentage of increase in built-up areas, which is a straightforward measure of urban growth rate, has also been calculated for each borough in KMC and for each time span.

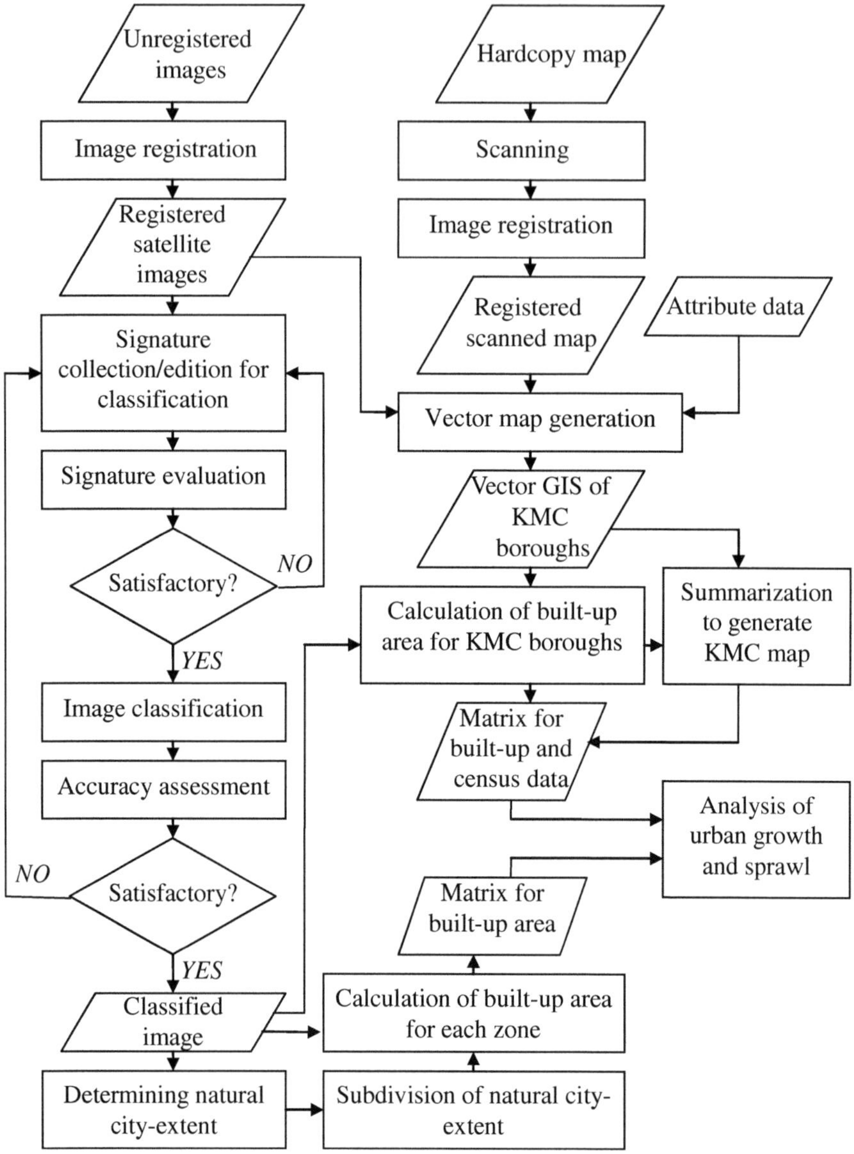

Fig. 3.6 Flowchart of the methodology

3.13.2 Proportion of Population and Proportion of Built-up Area

Sudhira et al. (2004) showed that the proportion of population in every region to the built-up area of that region is a measure of quantifying sprawl. However, this index does not consider the built-up areas within neighboring zones. Normally, built-up area within a zone is not only influenced by the population but also influenced by the built-up areas of its neighboring zones. If one zone saturates with built-up then the built-up grows into its neighboring zones. Therefore, it is better to relate the proportion of population and proportion of built-up area for each zone instead of regarding the proportion of population as being a function of built-up area of that zone.

The *proportion of population* and *proportion of built-up area* have been calculated by dividing the population and built-up area of the respective zone by the total population and total built-up area of KMC, respectively (Bhatta 2009a).

$$P_i = \frac{x_i}{\sum_{i=1}^{n} x_i} \tag{3.2}$$

where, P_i is the proportion of a phenomenon (variable) occurring in the ith zone, x_i is the observed value of the phenomenon in the ith zone, and n is the total number of zones.

In this approach, if we subtract the *proportion of population* from the *proportion of built-up area*, the results can vary within the range of -1 to 1, where 0 indicates the balanced condition.

This analysis has been done for the years 1980, 1990, and 2000. Since the population data of year 2010 or 2011 are yet to publish, the current status could not be determined. There are several equations available (e.g., Angel et al. 2005; Bhatta 2009a) to simulate and model the population data. However, these equations are generally reliable to interpolate the population data not to extrapolate. Population growth rate depends on many factors (such as birth rate, mortality rate, education, economic base, religion, and many more) that are beyond the scope of discussion in this title. However, these factors may highly vary from space to space and time to time. Therefore, reliability of extrapolated population data is not always guaranteed.

Although the preceding index of relating population and built-up is useful for intra-city comparative analysis; this index is a relative measure, and therefore, cannot be used for black-and-white characterization of sprawl for the entire city. This index was tested and reported in Bhatta (2009a), during this research, mainly to understand the disparity in distribution of built-up and population within the city.

One possibility of urban sprawl characterization is to consider the absolute measure of per capita built-up consumption, i.e., the ratio of *built-up area within a zone* to the *population within a zone*. However, this index is still weak in capturing the sprawl; because, what should be the per-capita consumption of land in a nonsprawling city is not well defined. Per-capita consumption of built-up area varies across cities. Population densities in cities of developing countries are on an average three times higher than the densities in developed country cities. The per-capita built-up consumption refers to utilization of all the land development initiatives such as the commercial, industrial, educational, and recreational establishments along with the residential establishments per person. Since most of the initiatives pave the way for the creation of jobs and subsequently help with earning livelihoods, the development of land can be seen as a direct consequence of this and hence one can conclude that the per-capita land consumption is inclusive of all the associated land development (Sudhira et al. 2004). Higher per-capita consumption of built-up area may indicate more living space for individuals as well as extended facilities to urbanites rather than sprawl (especially in highly dense cities like Kolkata).

3.13.3 Growth Rates of Population and Built-up Area

The growth rate of built-up areas is generally related to the growth rate of populations. Population growth forces the built-up areas to expand. Urban sprawl can also be identified hypothetically by careful examination of built-up area expansion rate and population growth rate (Sudhira et al. 2004; Bhatta 2009a). If the built-up growth rate exceeds the population growth rate it is generally considered as an indication of sprawl. If both are same, it is an ideal condition; and if the population growth rate exceeds the built-up growth rate it can be considered as population crowding. In this research, this analysis was performed for each borough as well as for entire KMC.

This approach, however, may be useful for the analysis of urban growth but cannot be accepted for black-and-white characterization of sprawl. It is often difficult to distinguish population change in a given jurisdiction as either the cause or effect of urban development; therefore, 'the population factor should not be used as a sole indicator of urban sprawl' (Ji et al. 2006). In the developing countries, population densities in cities are very high compared to developed countries. With the development of economic base, urban residents in developing countries generally seek some more living space and extended urban facilities (e.g., roads, schools, hospitals, theaters, etc.). Therefore, if the growth rate of built-up area exceeds the population growth rate, it may not indicate a sprawl. For example, the growth rate of population may be negative but the built-up area may remain same (built-up is generally irreversible; its growth cannot be negative). In this case, the preceding analysis will artificially show the area as being less compact. Further, a low built-up growth rate in an area does not guarantee a compact development.

3.13.4 Households and Built-up Area

In the previous section, it has been stated that the identification and quantification of sprawl may not be possible by comparing population and built-up data. Lambin et al. (2001) argued that in most cases, an increasing (or diminishing) number of built-up activities like housing and commercial constructions would be more effective to indicate sprawl as consequences of land consumption because usually construction activities, as compared to population change, reflect directly economic opportunities as the major driving force of land alteration. In a recent effort, the concept 'housing unit' was used as a proxy for population and combined with digital orthophoto data to generate urban sprawl metrics (Hasse and Lathrop 2003).

Built-up areas generally increase with the increase in the number of households. But how they differ among zones is an important issue that may help us to identify the compactness of urban growth and to evaluate the built-up density. A simple approach may be to relate the household density with built-up density in each zone (borough) individually. But these variables within a zone are never independent; rather, they are influenced by the neighbors of the respective zone and the entire city as well. Therefore, it would be better to relate the proportion of households in a zone to the total households of the city (*A*) with the proportion of built-up area within the respective zone to the total built-up area of the city (*B*) (Bhatta 2009a). These proportions have been calculated by using the Eq. 3.2 for each KMC borough. The relation between these two proportions (*proportion of household* minus *proportion of built-up*, *A*–*B*) clearly shows the compactness/dispersion of a zone. If we consider 0 as the average condition, then positive values show the compactness and negative values show dispersion (Bhatta 2009a).

However, the preceding index is useful for the intra-city analysis of built-up area and its relative compactness as a pattern. This index cannot be used to identify whether the entire city is becoming more compact or more dispersed. Therefore, we need to observe the change in difference of *household density* and *built-up density* through time for the entire KMC. The change ($\Delta\rho$) between time t_1 and t_2 can be calculated using the following formula:

$$\Delta\rho = \rho(t_1) - \rho(t_2) \tag{3.3}$$

where, ρ = *household density* minus *built-up density*.

Household density has been calculated as number of households per hectare, and built-up density has been calculated as percentage of built-up area to the total area.

Although this proposed approach of analysis is helpful to identify the relative compactness among different dates, it is inferior in capturing the sprawl in black-and-white characterization. For example, what should be the ideal value of ρ (*household density* minus *built-up density*) in this measurement scale is undefined. Furthermore, the units of household density and built-up density are different and they are independent. Therefore, a better approach for the black-and-white characterization of sprawl may be to consider the absolute growth rates of household

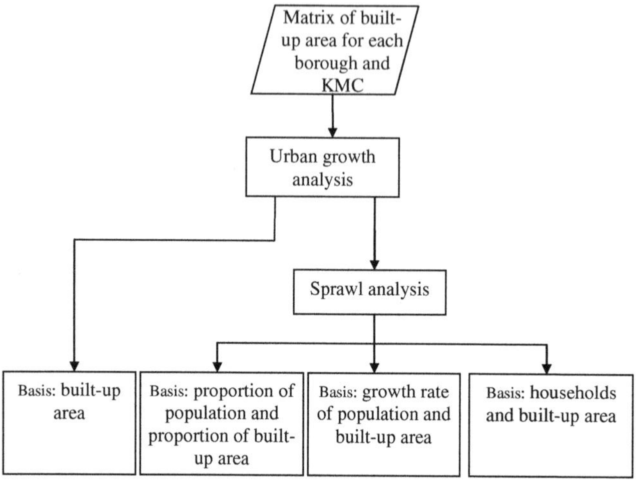

Fig. 3.7 Flowchart of analysis based on administrative boundary

and built-up within a zone. In this new approach, if we subtract the built-up growth rate from the household growth rate we can determine the compactness for a temporal span. If we consider 0 as the ideal condition, then positive values show the compactness and negative values show dispersion. This index has been applied on each borough and on entire KMC to quantify the sprawl in black-and-white characterization.

The entire analysis in consideration of administrative boundary can be summarized in a flowchart as shown in Fig. 3.7.

3.14 Analysis Based on Natural Boundary

The analysis of urban growth, especially urban sprawl, is primarily based on the spatial structure of built-up areas rather than any other phenomena. Since the spatial structure of built-up area is not dependent on the administrative boundaries or census tracts they should be analyzed in respect of natural boundary of the city. The city of Kolkata, in this research, has also been analyzed in consideration of the natural extent of the city. How the multi-temporal natural extents of the city have been delineated and how they have been subdivided, have been documented in Sects. 3.9 and 3.10.

The cities that are experiencing properly designed planning and restriction policies may have a predetermined urban growth boundary (UGB) for a specific date. In such cases, since the planning policies are regulated within the predetermined sharp boundaries, these UGBs may be considered instead of digitally determined natural boundary. However, since the city of Kolkata does not associate any such boundary, the natural extents have been considered in the analysis.

3.14.1 Area Index and Shape Index

As mentioned earlier in Chap. 2 (Sect. 2.5), there are dozens of metrics that are in use for the analysis of urban growth and sprawl. Limitations of them (e.g., dependency on spatial resolution, extensive data requirements, etc.) have also been discussed. However, let us consider two such 'simple' but widely practiced metrics—*area index* and *shape index* (McGarigal and Marks 1995; McGarigal et al. 2002; Herold et al. 2003; Angel et al. 2007; Jiang et al. 2007). The word 'simple' means less demanding in terms of data, computational power, and expertise.

Area index is the ratio of patch size and built-up area within the patch; hence, 'patch' is the city extent. This index is a relative measure of porosity, for which we may have a lowest value of 1 that indicates lowest level of porosity. Hence, 'porosity' is the nonbuilt-up areas within the built-up. This index can provide us with the information of relative porosity, whether it is increasing or decreasing with the change of time. However, since it is a relative measure it cannot be used for black-and-white characterization of sprawl. There is no reliable approach to define a threshold that can determine whether the city is sprawled or not. What should be the ideal value of porosity is also undefined. Scientists and researchers made several inferences and intuitively draw the conclusion whether the city sprawled or not. However, how a city administrator, who is generally not a scientist, will make these inferences is an obvious question. There remains a high possibility of misinterpretation of the results.

Another metric is shape index—edge length (perimeter) on a per unit area basis that facilitates to determine whether the shape is complex or simple (Fig. 3.8). A general assumption is higher shape index denotes higher complexity—means sprawl.

However, the review of the literature found that one such metric may contradict with other. Actually, most of the spatial metrics suffer from several limitations. We should carefully select these metrics while analyzing the urban growth and sprawl. 'Just because something can be computed does not mean that it should be computed' (Turner et al. 2001). Many metrics are highly correlated, for example, there is no difference between *total edge* (perimeter) and *shape index* (perimeter $\div$ area) if the patch sizes are similar. Furthermore, some of the metrics are not useful in some of the contexts of urban growth analysis. Urban growth is an obvious outcome of human civilization; it cannot be completely stopped or reversed. Therefore, with the increase of built-up area, no surprising, total edge index will also increase. That means an increasing total edge index is a general outcome of the urban growth process; it does not mean the city is becoming dispersed. It may happen that built-up growth rate is exceeding the growth rate of total edge index; it is an indication of decreasing porosity. In this instance, the city is becoming more compact rather than sprawled.

Coming back to the shape index, in many instances complex shape may also represent compact development and vice versa. If one interprets the images shown in Fig. 3.9, one can understand that the city extent is more complex in the left

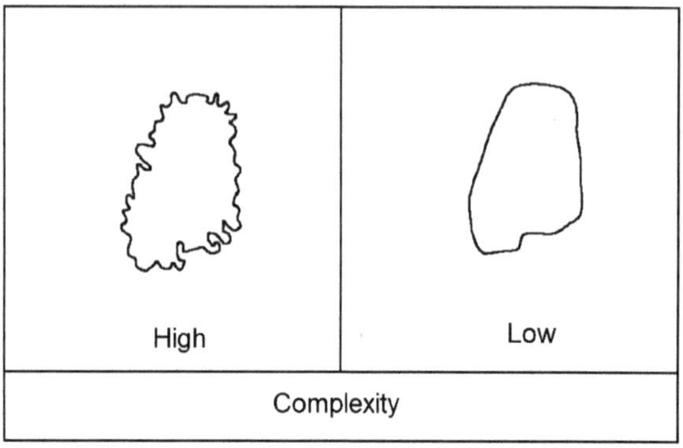

Fig. 3.8 Complex shapes have higher shape index (Bhatta 2010)

image but it is more compact than the right-hand side image. The city extent is more porous in the right image which is an indication of sprawl. Furthermore, shape index is also a relative measure and incapable for the black-and-white characterization of sprawl; there is no defined threshold to identify sprawl and nonsprawl. Therefore, it is better to avoid the perimeter phenomenon in the analysis of urban sprawl.

Owing to the limitations of the preceding indices, it is better not to draw any conclusion relying on them. Rather, we should choose or develop some metrics that can overcome these limitations and can identify the sprawl in black-and-white characterization.

3.14.2 City Extent and Built-up Area

To overcome the limitations stated in the preceding section, a new index has been developed during this research and reported in Bhatta (2009b). Urban sprawl can be identified by careful examination of the *area of the city extent* and *built-up area within the city extent*. This simple approach of determining the urban sprawl is—if the growth rate of the city extent exceeds the built-up growth rate, it is an indication of sprawl. Positive values in their differences (*city extent growth rate* minus *built-up growth rate*) indicate sprawl and negative values indicate compactness. The degree of sprawl and compactness can also be identified by their magnitudes. Actually, this index indirectly reveals the porosity within the city extent. Higher built-up growth with a lower growth in city extent is surely a compact development, and vice versa.

Built-up growth rate and the rate of growth in city extent, for the city of Kolkata, have been calculated in this research. Differences between them have also been calculated for the analysis of urban sprawl.

However, as with the area index and shape index, the major limitation of this index is the consideration of the entire city in the analysis. It does not consider the variations of the built-up areas in different parts of the city.

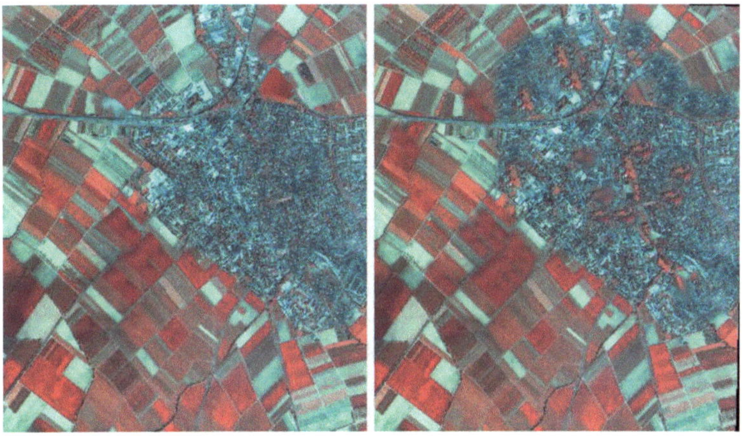

Fig. 3.9 Shape index may not relate urban sprawl (Courtesy: National Remote Sensing Centre, India)

3.14.3 Shannon's Entropy and Built-up

Entropy method can overcome the limitations stated in the preceding section. As mentioned in Sect. 2.5 (Chap. 2), entropy method is the most widely used metric for urban sprawl analysis. The review of the literature also found that the entropy method is the most reliable metric among the available sprawl measurement indices. In this research, the entropy method has been applied to serve two purposes: Firstly, to overcome the limitations of demonstrated indices and to obtain a reliable result by using the most widely used metric; secondly, to evaluate the results obtained by applying other indices, especially the newly proposed and modified indices.

Shannon's entropy (*H*) can be used to measure the degree of spatial concentration or dispersion of a geographical variable (x_i) among zones (Theil 1967; Thomas 1981). Entropy is calculated by:

$$H_n = \sum_{i=1}^{n} P_i \log_e\left(\frac{1}{P_i}\right) \tag{3.4}$$

where, P_i is the proportion of a phenomenon occurring in the *i*th zone $\left(P_i = \frac{x_i}{\sum_{i=1}^{n} x_i}\right)$, x_i is the observed value of the phenomenon in the *i*th zone, and *n* is the total number of zones. The value of entropy ranges from 0 to $\log_e(n)$. A value of 0 indicates that the distribution of built-up areas is very compact, while values closer to $\log_e(n)$ reveal that the distribution of built-up areas is dispersed. Higher values of entropy indicate the occurrence of sprawl. Half-way mark of $\log_e(n)$ is generally considered as threshold. If the entropy value crosses this threshold the city is considered as sprawled.

Relative entropy can be used to scale the entropy value into a value that ranges from 0 to 1. As Thomas (1981) demonstrated, relative entropy (H') for n number of zones can be calculated as:

$$H'_n = \frac{H_n}{\log_e(n)} \tag{3.5}$$

In this instance 0.5 is considered as threshold. Values higher than this generally considered as sprawl.

In the current research, built-up areas within each zone (zones are shown in Fig. 3.5) and for each date (1980, 1990, 2000, and 2010) have been calculated. If we calculate the entropy from these built-up data, we can get the entropy values for the four temporal dates. This model is robust, because it can identify the sprawl as a pattern for each date in black-and-white characterization. The threshold, that can determine whether the city is sprawled or compact, can also be determined mathematically.

However, the preceding approach cannot analyze the sprawl as a process. Yeh and Li (2001) argue that because entropy can be used to measure the distribution of a geographical phenomenon, the measurement of the difference of entropies (or relative entropies) between time t_1 and t_2 can be used to indicate the magnitude of change in urban sprawl as follows:

$$\Delta H_n = H_n(t_2) - H_n(t_1) \tag{3.6}$$

$$\Delta H'_n = H'_n(t_2) - H'_n(t_1) \tag{3.7}$$

The change in entropy can be used to identify whether the land development is becoming more dispersed (sprawled). However, as mentioned in Sect. 2.5 (Chap. 2), entropy is suffered from modifiable areal unit problem (MAUP). Relative entropy can mitigate the scale effect of MAUP. But, zone effect can only be overcome by decomposition of entropy.

Following the research of Thomas (1981), let us now apply the entropy decomposition theorem in this research to overcome the problems associated with zone effect. As a preliminary to decomposition analysis, zoning systems must be delimited for the different scales of analysis. We have four zones (as shown earlier in Fig. 3.5) that can be summarized into two zones, first one by combining Zones 1 and 2, and second one by combining Zones 3 and 4. Let these zones are *Delimited Zones* 1 and 2. The proportion of built-up area can then be calculated as:

$$P_j = \sum_{i \in j}^{n_j} P_j(i) \tag{3.8}$$

where, $i \in j$ denotes the value of i which is the first element ($\in$) of set j, and n_j is the value of i which forms the last element of set j. P_j is the proportion of built-up in a delimited zone and $P_j(i)$ is the proportion in a zone before delimitation.

The entropy decomposition theorem states that the entropy should be calculated in terms of both the proportions $P_j(i)$ and P_j. Hence the formula of entropy can be written as (Thomas 1981):

$$H_n = \sum_{j=1}^{k} P_j \log_e\left(\frac{1}{P_j}\right) + \sum_{j=1}^{k}\left[P_j \sum_{i\in j}^{n_j}\left(\frac{P_j(i)}{P_j}\right)\log_e\left(\frac{P_j}{P_j(i)}\right)\right] \tag{3.9}$$

where, k is the total number of delimited zones (i.e., 2), and other symbols are as stated earlier.

Equation 3.9 is composed of two expressions on either side of the addition (+) sign. The expression on the left-hand side is termed the *between region entropy* and is denoted by H_k, while the expression in the right-hand side is termed the *within region average entropy* and is denoted by $H_{n/k}$. This terminology allows that entropy decomposition equation (Eq. 3.9) to be written simply as:

$$H_n = H_k + H_{n/k} \tag{3.10}$$

Notice that, taken together, the entropies associated with maximum dispersion satisfy the decomposition theorem (Eq. *3.10*), because

$$\log_e(n) = \log_e(k) + \log_e(n/k) \tag{3.11}$$

Therefore, the relative between-region entropy has been calculated as:

$$H'_k = \frac{H_k}{\log_e(k)} \tag{3.12}$$

Relative within-region average entropy has been calculated as:

$$H'_{n/k} = \frac{H_{n/k}}{\log_e(n/k)} \tag{3.13}$$

These relative entropies should be interpreted separately since they can reveal opposite conclusions (as presented by Thomas 1981).

3.14.4 Degree of Freedom in Urban Growth

The preceding analysis shows that the entropy method is robust for the analysis and measurement of sprawl. However, as mentioned in Sect. 2.5 (Chap. 2), analysis or quantification of sprawl is not adequate in the analysis of urban growth. Whether the observed growth meets the planed (expected) growth and to what extent they meet are also necessary to evaluate.

However, for the cities without having any predetermined planned or expected values, such as Kolkata, it is not possible to execute this analysis. Therefore, the expected values, for the city of Kolkata, have been computed statistically. One of the objectives of this research is also to propose new model towards the analysis of the urban growth, which can be applied on other cities as well. Analysis of the urban growth is more demanded by the administrators of the cities that are consistently experiencing properly planned policies. Therefore, to demonstrate the models, the expected values were necessary to derive statistically. However, these values have been used only to demonstrate the models, not to make any comment on the city of Kolkata.

Built-up areas in each zone can be calculated and presented in a table where columns will indicate the zones and rows will indicate years under analysis (refer Table 4.12 in Chap. 4). Let this table be called matrix M, with elements M_{ij}, where $i = 1, 2, \ldots\ldots, n$ (specific date under analysis, rows of the table) and $j = 1, 2, \ldots\ldots, m$ (specific zone, columns of the table). The expected built-up area for each variable can be calculated by the products of marginal totals, divided by the grand total. Therefore, the expected growth M_{ij}^{E} for the i-th row and j-th column is (Healey and Prus 2009; Almeida et al. 2005):

$$M_{ij}^{\mathrm{E}} = \frac{M_i^{\mathrm{S}} \times M_j^{\mathrm{S}}}{M^{\mathrm{G}}} \tag{3.14}$$

where, M_i^{S} = row total, M_j^{S} = column total, and M^{G} = grand total = $\sum_{i=1}^{n}\sum_{j=1}^{m} M_{ij}$.

In probability theory and statistics, the *expected value* (or *expectation value*, or *mathematical expectation*, or *theoretical expectation*) of a random variable is the integral of the random variable with respect to its probability measure. For discrete random variables this is equivalent to the probability weighted sum of the possible values. For continuous random variables with a density function it is the probability density-weighted integral of the possible values (Healey and Prus 2009).

The term 'expected value' may be misleading; it is not the 'most probable value'. The expected value, in general, is not a typical value that the random variable can take on. It is often helpful to interpret the expected value of a random variable as the long-run average value of the variable over many independent repetitions of an experiment. The expected value may be intuitively understood by the *law of large numbers*: the expected value, when it exists, is almost surely the limit of the sample mean as sample size grows to infinity (Healey and Prus 2009). The value may not be 'expected' in the general sense—the expected value itself may be unlikely or even impossible (e.g., expected built-up area within a zone may exceed the total area of the zone). This limitation was completely overlooked by Almeida et al. (2005) and Bhatta et al. (2010b).

Since the theoretical expected values may be misleading and may not represent the actual expectations, no conclusions have been made based on these values. Rather, these theoretical expectations have been derived to demonstrate some analytical models. For the city of Kolkata, without having any predetermined expected values, these models could not be demonstrated otherwise.

The expected and observed values can be used to determine the *degree of freedom*. In statistics, the number of degrees of freedom is the number of values in the final calculation of a statistic that are free to vary. The term is most often used in the context of linear models (*linear regression*, *analysis of variance*), where certain random vectors are constrained to lie in linear subspaces, and the degrees of freedom are the dimension of the subspace. The degrees of freedom are also commonly associated with the squared lengths (or 'sum of squares') of such vectors, and the parameters of Chi-squared and other distributions that arise in associated statistical testing problems (Walker 1940).

In probability theory and statistics, the *Chi-square statistic* (also called *Chi-squared test* or *Chi-squared distribution*) with k degrees of freedom is the distribution of a sum of the squares of k independent standard normal random variables. It is one of the most widely used probability distributions in inferential statistics, e.g., in hypothesis testing, or in construction of confidence intervals (Mood et al. 1974). The best-known situations in which the Chi-square distribution is used are the common Chi-square tests for goodness of fit of an observed distribution to a theoretical one, and of the independence of two criteria of classification of qualitative data (Mood et al. 1974).

Pearson's Chi-square test is the best-known of several Chi-square tests whose results are evaluated by reference to the Chi-square distribution (Plackett 1983). Its properties were first investigated by Karl Pearson. It tests a null hypothesis stating that the frequency distribution of certain events observed in a sample is consistent with a particular theoretical distribution.

Pearson's Chi-square statistic takes into account the checking of freedom amongst pairs of variables chosen to explain the same category of land cover change (Almeida et al. 2005; Bhatta et al. 2010b). Therefore, to determine the 'degree of freedom' in urban growth, Chi-square test has been performed with the Pearson's Chi-square expression:

$$\chi^2 = \sum_{j=1}^{m} \left(\frac{M_j - M_j^{\mathrm{E}}}{M_j^{\mathrm{E}}} \right)^2 \tag{3.15}$$

where, χ^2 = degree of freedom, M_j = observed built-up area in j-th column, M_j^{E} = expected built-up area in j-th column.

The Chi-square (χ^2) has a lower limit of 0, when the observed value exactly equals the expected value. Higher magnitude shows higher degree of freedom. The measurement of the difference of Chi-square values ($\Delta\chi^2$) between time t_1 and t_2 can be used to indicate the magnitude of change in degree of freedom as follows:

$$\Delta\chi^2 = \chi^2(t_2) - \chi^2(t_1) \tag{3.16}$$

This equation helps to analyze the degree of freedom as a process.

This method of analysis is a new approach for the analysis of freedom in urban growth. Bonham-Carter (1994) and Almeida et al. (2005) have used Chi-square for determining the overall freedom (considering all values in the matrix for computation) that results in a single value of freedom for all dates and all zones but the current research shows how this model can be used to analyze the urban growth as a pattern as well as process. However, it is worth mentioning that higher degree of freedom cannot be considered as sprawl, instead it should be considered as disparity in expected and observed growth as a pattern and/or process.

Although Bhatta et al. (2010b) demonstrated the approach to determine the freedom as process, pattern, and 'overall', their method is not acceptable (explained in Sect. 3.14.5).

3.14.5 Degree of Goodness in Urban Growth

As mentioned in Sect. 2.5 (Chap. 2) entropy (degree of sprawl) and Chi-square (degree of freedom) are different measures and one may not relate other. Therefore, it necessitates determining the 'degree of goodness' of the urban growth. The *degree of goodness* actually refers to the degree to which observed growth relates the planned growth and the magnitude of compactness. This can be calculated for each date under analysis by using the following formula:

$$G = \log_e\left(\frac{1}{\chi^2 H'_n}\right) \tag{3.17}$$

where,

G = degree of goodness

χ^2 = degree of freedom (as calculated by the Eq. 3.15)

H'_n= relative entropy (as calculated by the Eq. 3.5)

Degree of goodness is a straightforward measure; positive values indicate 'goodness' whereas negative values indicate 'badness'; a value of 0 is an indication of average condition. If we calculate the difference in goodness (ΔG) between time t_1 and t_2, we can understand the goodness as a process.

$$\Delta G = G(t_2) - G(t_1) \tag{3.18}$$

Degree of goodness can be a robust measure of the urban growth analysis; because, it can embrace the sprawl measures and the measures that estimate the degree to which the urban growth meets the planning (or restriction) policies. It can analyze the urban growth in black-and-white characterization; for example, whether the urban growth is good or bad can be identified by their signs, and the magnitude of goodness or badness can also be identified.

The concept of degree of goodness has been developed during this research and initially reported in Bhatta et al. (2010b). However, it was applied in a different approach in the said paper. The entropy had been calculated to measure the sprawl

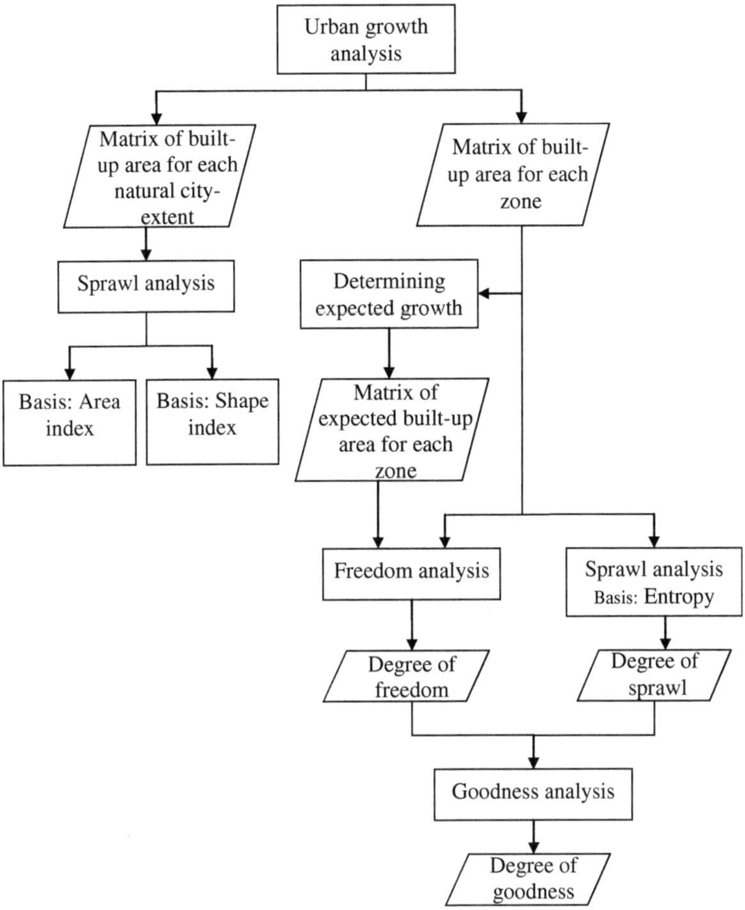

Fig. 3.10 Flowchart of analysis based on natural extents of the city

as a pattern, process, and 'overall' (*overall* combines both pattern and process). Entropies had been calculated from a built-up growth matrix (not the built-up area matrix as demonstrated in this title). Degree of freedom and degree of goodness have also been calculated based on the built-up growth matrix as a pattern, process, and 'overall'. Important to mention, the sprawl, freedom, and goodness as a process had been quantified in a different approach than the methods described in this book (columns of built-up growth matrix formed the basis of analysis). Later it has been realized that the sprawl phenomenon relates to the dispersion of built-up area; therefore, the use of built-up growth (instead of built-up area) to determine the entropy/freedom (and thus the goodness) is a questionable approach (reported in Bhatta et al. (2010a)). This title, therefore, considers the built-up area matrix instead of built-up growth matrix.

Finally, to summarize the entire analysis in consideration of natural extents of the city, a flowchart is being presented in Fig. 3.10.

References

Almeida CM, Monteiro AMV, Mara G, Soares-Filho BS, Cerqueira GC, Pennachin CSL, Batty M (2005) GIS and remote sensing as tools for the simulation of urban land-use change. Int J Remote Sens 26(4):759–774

Anderson JR, Hardy EE, Roach JT, Witmer RE (1976) A land use and land cover classification system for use with remote sensor data. US geological survey professional paper 964. United States Government Printing Office, Washington. http://landcover.usgs.gov/pdf/anderson.pdf.

Angel S, Parent J, Civco D (2007) Urban sprawl metrics: an analysis of global urban expansion using GIS. Proceedings of ASPRS 2007 annual conference, Tampa, Florida May 7–11. http://clear.uconn.edu/publications/research/tech_papers/Angel_et_al_ASPRS2007.pdf

Angel S, Sheppard SC, Civco DL (2005) The dynamics of global Urban expansion. Transport and Urban Development Department, The World Bank, Washington, p 200. http://www.citiesalliance.org/doc/resources/upgrading/urbanexpansion/worldbankreportsept2005.pdf

Arnold CL Jr, Gibbons JC (1996) Impervious surface coverage—the emergence of a key environmental indicator. J Am Plan Assoc 62(2):243–258

Barnes KB, Morgan III JM, Roberge MC, (2000) Impervious surfaces and the quality of natural and built environments. A white paper prepared for the project chesapeake bay from space. www.Chesapeakebayfromspace.net.

Barnes KB, Morgan JM III, Roberge MC, Lowe S (2001) Sprawl development: its patterns, consequences, and measurement. A white paper, Towson University. http://chesapeake.towson.edu/landscape/urbansprawl/download/Sprawl_white_paper.pdf

Barnsley MJ, Barr SL (2000) Monitoring urban land use by earth observation. Surv Geophys 21:269–289

Bhatta B (2008) Remote Sensing and GIS. Oxford University Press, New York and New Delhi, p 720

Bhatta B (2009a) Analysis of Urban growth pattern using remote sensing and GIS : a case study of Kolkata, India. Int J Remote Sens 30(18):4733–4746

Bhatta B (2009b) Modelling of Urban growth boundary using geoinformatics. Int J Digit Earth 2(4):359–381

Bhatta B (2010) Analysis of Urban growth and sprawl from remote sensing data. Springer-Verlag, Heidelberg, Berlin, p 172

Bhatta B, Saraswati S, Bandyopadhyay D (2010a) Urban sprawl measurement from remote sensing data. Appl Geogr 30(4):731–740

Bhatta B, Saraswati S, Bandyopadhyay D (2010b) Quantifying the degree-of-freedom, degree-of-sprawl, and degree-of-goodness of Urban growth from remote sensing data. Appl Geogr 30(1):96–111

Bonham-Carter G (1994) Geographic information systems for geoscientists: modelling with Gis. Computer methods in the geosciences, vol 13. Elsevier, New York, p 398

Census (1981) District census handbook, series 23, parts XIII A & B. Directorate of census operations, West Bengal

Census (1991) District census handbook, series 26, parts XII A & B. Directorate of census operations, West Bengal

Census (2001) Table-4, table-12, table-13, and table-B21. Directorate of census operations, West Bengal

Epstein J, Payne K, Kramer E (2002) Techniques for mapping suburban sprawl. Photogramm Eng Remote Sens 63(9):913–918

ERDAS (2008) ERDAS field guide, vol. 1 & 2. Leica Geosystems Geospatial Imaging, Norcross

Hasse J, Lathrop RG (2003) A housing-unit-level approach to characterizing residential sprawl. Photogramm Eng Remote Sens 69(9):1021–1030

Healey JF, Prus, SG (2009) Statistics: a tool for social research. Cengage learning

Herold M. Goldstein N, Clarke KC (2003) The spatio-temporal form of urban growth: measurement, analysis and modeling. Remote Sens Environ 86(3):286–302

Jensen JR (2005) Introductory digital image processing: a remote sensing perspective. Prentice-Hall, Upper Saddle River

Ji W, Ma J, Twibell RW, Underhill K (2006) Characterizing urban sprawl using multi-stage remote sensing images and landscape metrics. *Comput Environ urban syst* 30:861–879

Jiang F, Liu S, Yuan H, Zhang Q (2007) Measuring urban sprawl in Beijing with geo-spatial indices. J Geogr Sci doi:10.1007/s11442-007-0469-z

Lambin EF, Turner BL, Geist HJ, Agbola SB, Angelsen A, Bruce JW et al. (2001) The causes of land-use and land-cover change: moving beyond the myths. Global Environ Chang 11:261–269

Liu H, Zhou Q (2004) Accuracy analysis of remote sensing change detection by rule-based rationality evaluation with post-classification comparison. Int J Remote Sens 25(5):1037–1050

McGarigal K, Marks BJ (1995) FRAGSTATS: spatial pattern analysis program for quantifying landscape structure. USDA forest service general technical report PNW-351

McGarigal K, Cushman SA, Neel MC Ene E (2002) FRAGSTATS: spatial pattern analysis program for categorical maps. www.umass.edu/landeco/research/fragstats/fragstats.html.

Mood A, Franklin AG, Duane CB (1974) Introduction to the theory of statistics 3rd edn. McGraw-Hill, p 241–246

Neer JT (1999) High resolution imaging from space—a commercial perspective on a changing landscape. Int Arch Photogramm Remote Sens 32(7C2):132–143

Plackett RL (1983) Karl pearson and the chi-squared test. Int Statis Rev 51(1):59–72

Schueler TR (1994) The importance of imperviousness. Watershed Prot Tech 3(1):100–111

Sudhira HS, Ramachandra TV, Jagdish KS (2004) Urban sprawl: metrics, dynamics and modelling using GIS. Int J Appl Earth Obs Geoinformation 5:29–39

Theil H (1967) Economics and information theory. North-Holland, Amsterdam, p. 488

Thomas RW (1981) Information statistics in geography, Geo abstracts. University of East Anglia, Norwich, p. 42

Torrens PM, Alberti M (2000) Measuring sprawl. Working paper no. 27, Centre for advanced spatial analysis, University College, London. http://www.casa.ac.uk/working papers/.

Turner MG, Gardner RH, O'Neill RV (2001) Landscape Ecology in Theory and Practice: Pattern and Process. Springer, New York, p. 401

Walker HM (1940) Degrees of freedom. J Educational Psychol 31(4):253–269

Wilson EH, Hurd JD, Civco DL, Prisloe S, Arnold C (2003) Development of a geospatial model to quantify, describe and map urban growth. Remote Sens Environ 86(3):275–285

Yeh AGO, Li X (2001) Measurement and monitoring of urban sprawl in a rapidly growing region using entropy. Photogramm Eng Remote Sens 67(1):83–90

Chapter 4
Results and Discussion

Abstract This chapter displays the results obtained from the digital classification of remote sensing data as well as the results obtained from different analyses. The study concludes that, for the analysis of urban growth and sprawl, the natural boundary of the city is preferred compared to the administrative boundary. However, whether the administrative or the natural boundaries are in consideration, it has been found that the city of Kolkata was always sprawling. The results have also shown that the rate of urban growth in Kolkata is generally higher than the growth rate of the population as well as the growth rate of households. Among the metrics and models used to quantify the sprawl, Shannon's entropy has been established as the most preferred. This chapter also focuses the discussions on the observed results. It also makes arguments on the methodology in terms of their merits and limitations, and constructs logical statements.

Keywords Urban growth · Sprawl analysis · Remote sensing · GIS · City growth evolution · Built up area concentration · Population growth rate

4.1 Classified Imagery and Change Map

Digital classification of the satellite images into built-up (with other impervious) and nonbuilt-up areas, for four temporal dates (1980, 1990, 2000, and 2010), has resulted in the creation of abstracted and highly simplified visual images of the study area as shown in Fig. 4.1, which are important evidences of urban extents and growth patterns. By examining the classified images, the growth patterns of the city in different areas, the infill of the open spaces between already built-up areas, the extents of urban area, etc. can be identified intuitively.

An overlay of the classified images, as shown in Fig. 4.2, can provide a *change map* that can illustrate the pattern and process of urban growth. However, to describe these different patterns intelligently, to understand how they change over

B. Bhatta, *Urban Growth Analysis and Remote Sensing*, SpringerBriefs in Geography,
DOI: 10.1007/978-94-007-4698-5_4,

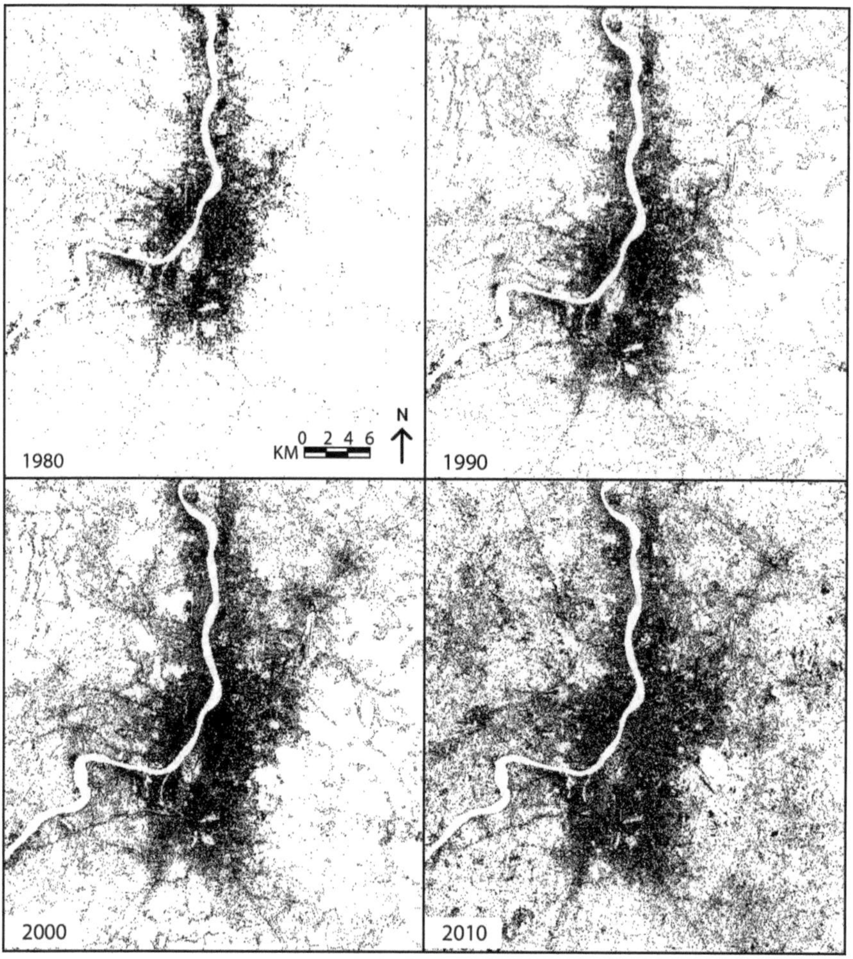

Fig. 4.1 Classified images from different dates showing built-up areas in *black*

time, to compare each zone or borough with others numerically, or to explain the variations among these patterns statistically, we need to select quantitative measures that summarize one or another of their properties.

4.2 Results From Analysis: Based on Administrative Boundary

This section furnishes the results obtained from the analyses based on administrative boundary.

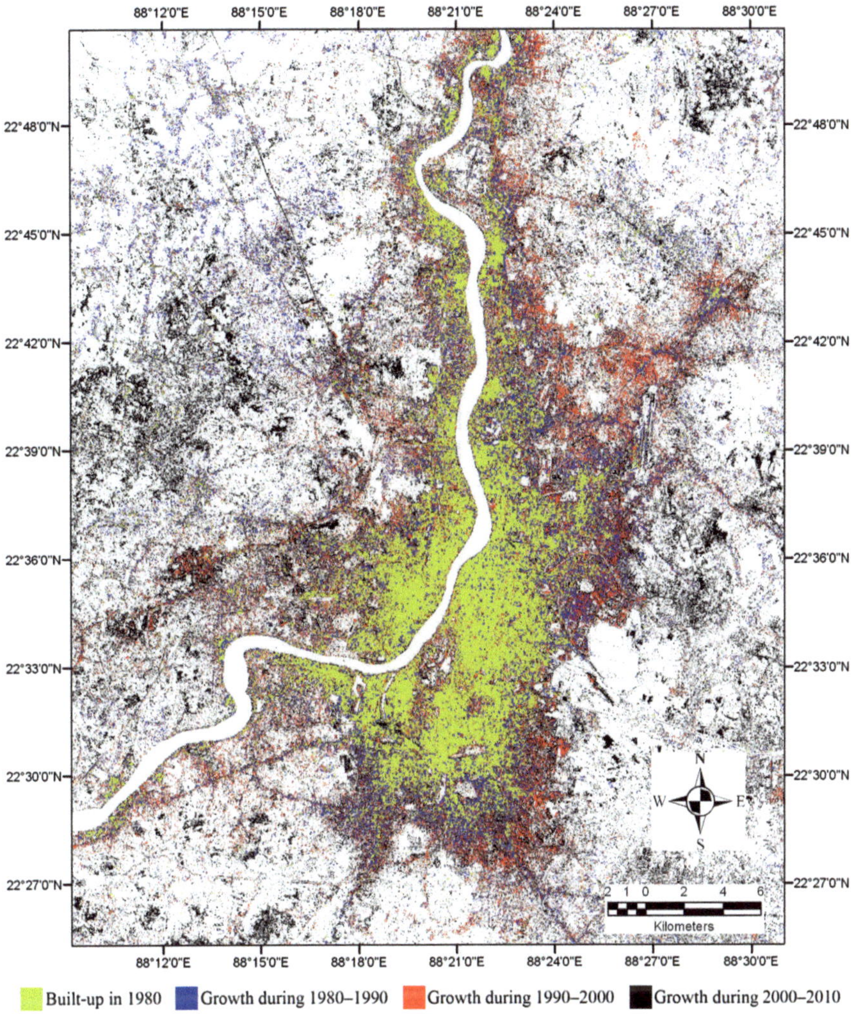

Fig. 4.2 Urban change map shows built-up growth in the Kolkata metropolitan

4.2.1 Built-up Area and Urban Growth

The built-up areas for each borough and for each date under analysis are furnished in Table 4.1 that directly shows the status of the built-up areas in the city. Table 4.1, Figs. 4.1 and 4.2 collectively inform us about the built-up status of the Kolkata Municipal Corporation (KMC).

If we prepare a chart from Table 4.1, it can show the trend of built-up growth history for each year under analysis and each borough, as well as the entire KMC. The chart shown in Fig. 4.3 reveals that the built-up areas in each borough (and in

Table 4.1 Percentage of built-up areas in KMC boroughs

Year	Borough															KMC
	1	2	3	4	5	6	7	8	9	10	11	12	13	14	15	
1980	66.78	85.56	61.13	87.47	79.91	82.08	31.97	72.15	57.93	41.70	5.17	5.51	23.41	13.33	40.20	38.54
1990	73.34	92.23	69.95	92.93	81.61	82.70	35.30	74.33	62.99	57.56	14.94	10.88	34.57	26.06	44.11	45.75
2000	83.66	96.72	80.61	96.70	91.86	91.45	45.01	84.65	74.28	71.75	43.71	29.56	49.25	39.73	58.82	59.17
2010	85.29	97.19	81.58	96.73	91.88	91.65	50.51	85.33	75.83	79.81	58.23	47.12	59.41	51.04	60.48	66.09

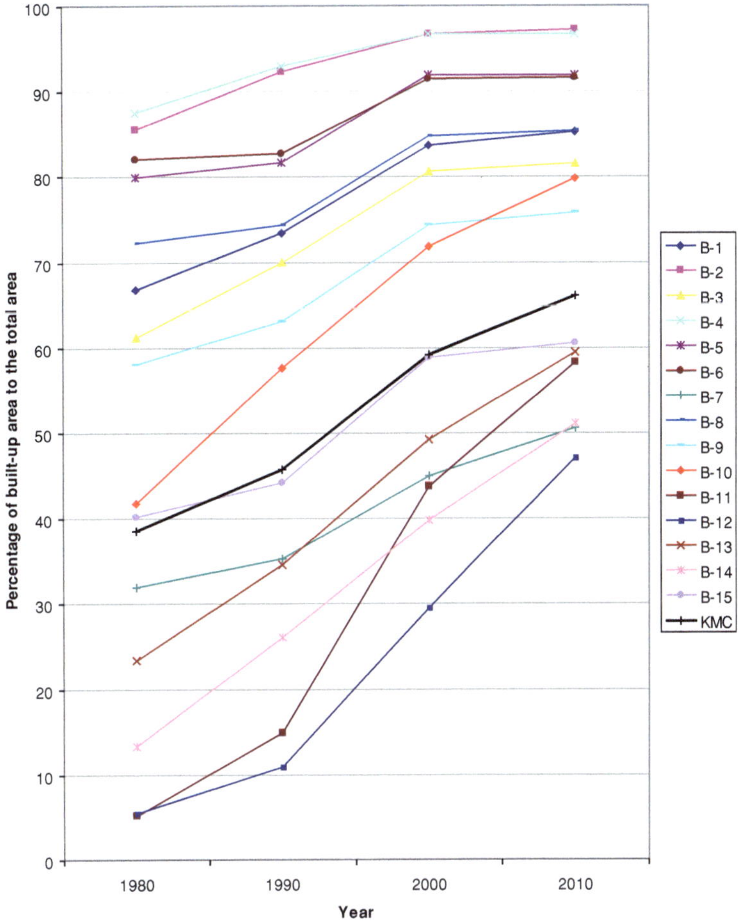

Fig. 4.3 Concentration of built-up areas (boroughs have been designated as B-1, B-2, B-3, and so on)

entire KMC as well) are continuously increasing. However, built-up areas are not distributed evenly. Northern and central Kolkata, the oldest urban areas, show a high concentration of built-up area, whereas the newly developed areas show less amount of built-up. This figure also shows that previously, this variation in built-up distribution was very high and this variation is continually declining. Spatial and temporal disparity in built-up distribution is perhaps a general outcome of unplanned urban growth.

The percentage of increase in built-up areas for each time span is presented in Table 4.2. This table provides with us information on the urban growth in different areas within the KMC. This table shows highly variable growth rates. For example, borough 11 has shown 192.55 % growth during 1990–2000, whereas borough 5

Table 4.2 Observed Built-up growth rate for each borough and KMC

Time span	Borough															KMC
	1	2	3	4	5	6	7	8	9	10	11	12	13	14	15	
1980–1990	9.82	7.79	14.42	6.24	2.12	0.76	10.41	3.02	8.74	38.05	189.05	97.65	47.68	95.47	9.72	18.71
1990–2000	14.08	4.87	15.25	4.06	12.57	10.57	27.50	13.89	17.93	24.65	192.55	171.59	42.45	52.43	33.35	29.33
2000–2010	1.95	0.49	1.20	0.03	0.02	0.22	12.22	0.80	2.09	11.23	33.22	59.40	20.63	28.47	2.82	11.71

Table 4.3 Proportion of built-up area minus proportion of population for KMC boroughs

Year[a]	Borough														
	1	2	3	4	5	6	7	8	9	10	11	12	13	14	15
1980	0.01	−0.02	−0.01	−0.03	−0.01	−0.01	0.03	0.02	0.06	0.01	–	–	–	–	–
1990	0.01	−0.02	0.00	−0.03	−0.01	−0.01	0.01	0.01	0.06	0.02	−0.02	−0.01	0.01	0.01	−0.01
2000	0.00	−0.02	−0.01	−0.03	−0.01	−0.02	0.01	0.01	0.05	0.01	0.00	0.01	0.01	0.01	−0.01

[a] Census data of 1981 are not available for boroughs 11–15. Total population of 1981 includes South Suburban, Gardenreach, and Jadavpur (i.e., 4126139 persons). Population data of 2011 are not available

has shown 0.02 % growth during 2000–2010. This variation in growth rate is mainly because of variation in built-up concentration. Some of the areas are saturated with built-up and some of the areas are newly developing. Figure 4.3 shows that high growth in built-up areas can be observed in the areas of low built-up concentration; and as the built-up concentration goes to saturation stage the growth rate declines. This is a general outcome of the urban growth process. This highly varying built-up growth rates also indicate that the KMC area has highly variable developable lands in different areas. One important thing has been observed in every case, the built-up growth rates are low in the time span 2000–2010 compared to the previous decades.

The results presented in this section help us to understand the urban growth as a pattern as well as a process, for the entire KMC area, as well as each subdivided area (borough). However, these results cannot provide any insight into urban sprawl.

4.2.2 Proportion of Population and Proportion of Built-up Area

As stated in Sect. 3.13.2 (Chap. 3), *Proportion of built-up area* minus *proportion of population* can aid the analysis of sprawl. The results can vary within the range of −1 to 1 as shown in Table 4.3, where 0 indicates the balanced condition. Higher negative values indicate population crowding, which may cause serious environmental problems, traffic congestion, and minimal social (and other) facilities. Higher positive values indicate relatively higher per capita consumption of built-up area. The results show that most of the boroughs are in approximate balanced conditions.

This index is a relative measure for intra-city analysis. Therefore, the results cannot help us to understand whether the entire KMC area is becoming more compact or dispersed.

4.2.3 Growth Rates of Population and Built-up Area

Figure 4.4 compares the growth rates of population and built-up for each borough and entire KMC. It has been noticed that some of the boroughs (borough 1, 2, 3, 4, 5, 8 and 9) have registered negative population growth rates. These areas, mainly in the central and northern parts of Kolkata, are the oldest urban areas of the city. Recently, many residential buildings in these areas are being converted into business or commercial buildings. Negative population growth rates may be a consequence of that conversion. Recently, people are showing interest to settle down at the periphery of the city, since peripheral wards are showing high population, as well as built-up growth rates.

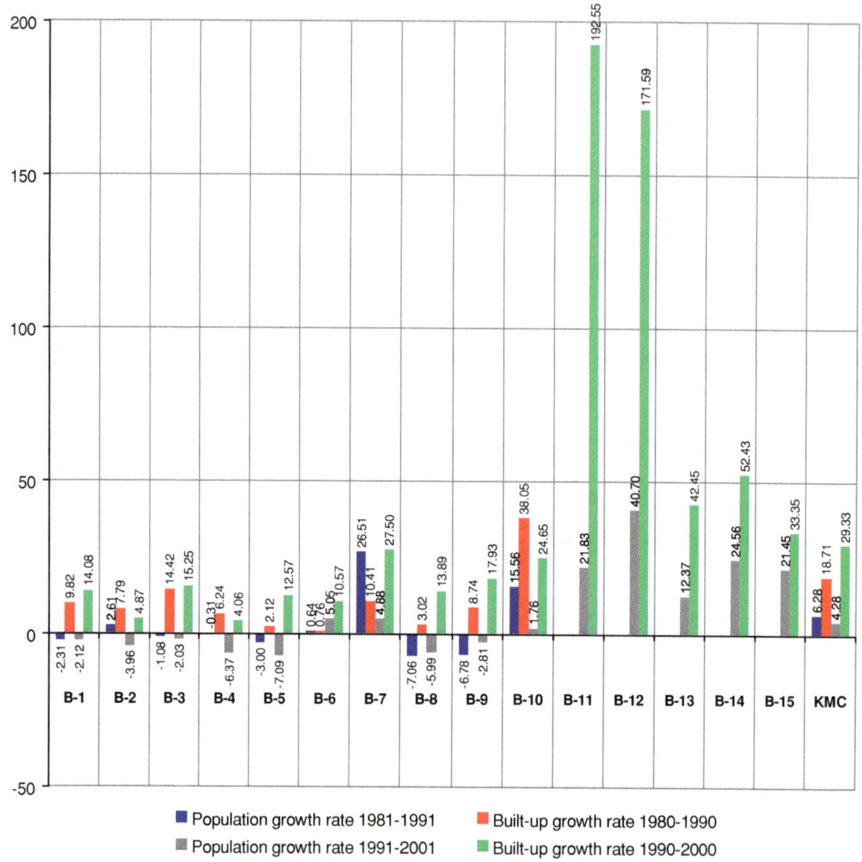

Fig. 4.4 Growth rate of population and built-up (boroughs have been designated as B-1, B-2, B-3, and so on)

The current research does not make any comment by extrapolating the population data of 2011. However, the census data of years 1981, 1991, and 2001 show that; in KMC, the population growth rates were always lower than the built-up growth rates. This clearly indicates increased per capita consumption of built-up area; i.e., the urbanites were enjoying increased urban facilities like more living space, more pavements, and more community level infrastructures. This is an indication of less compact urban growth.

This method results in a measure of compactness for the entire KMC as well as each subdivided part. However, as explained in Sect. 3.13.3 (Chap. 3), conclusions on sprawl cannot be made depending on this single observation, rather we also need to consider the housing units in the analysis.

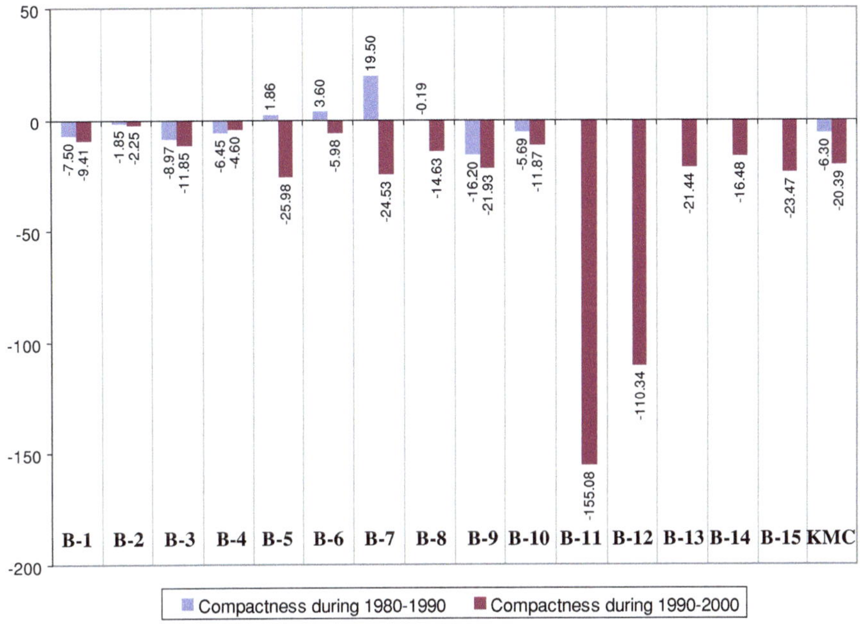

Fig. 4.5 Compactness for different temporal spans

4.2.4 Households and Built-up Area

Section 3.13.4 (Chap. 3) explains how the household data can be integrated with the built-up data for the analysis of sprawl. A new approach has been proposed: if we subtract the built-up growth rate from the household growth rate, we can determine the compactness for a temporal span. Figure 4.5 shows the compactness of each borough and the entire KMC for the time spans 1980–1990 and 1990–2000. Year 2010 could not be considered under analysis owing to the lack of census data. In Fig. 4.5, if we consider 0 as the ideal condition, then positive values show the compactness and negative values show dispersion.

This analysis also reveals similar conclusions that the KMC is becoming more dispersed. If we consider individual boroughs, the results shown in Fig. 4.5 support the results shown in Fig. 4.4. As one can see, for example, borough 11 has the highest sprawl and the next highest is borough 12, in both the analyses. Borough 7 was compact during 1980–1990; and in the time span 1990–2000 it was sprawling. Also, in case of other boroughs, both analyses reveal similar results. Therefore, the results can be accepted.

The main advantage of this index is perhaps the inclusion of vertical construction in the analysis. For example, let the built-up area remain unchanged and the number of households increase, then there must be two reasons: (1) one housing unit has been divided into smaller housing units (partition) and/or

(2) multistoried development has occurred in place of single-storied buildings. The first case is rare, because the city is highly dense and the population is continuously increasing. Therefore, logically, the second case is the obvious reason. Thus, the preceding index indirectly embraces the vertical construction.

Although this approach is useful for capturing the insights of urban growth and compactness as a pattern as well as a process, it is also not free from all nuisances. Figures of built-up growth generally consider all developmental initiatives (commercial, industrial, recreational, educational, transportation network, as well as housing units). Therefore, how many built-up areas are actually occupied by residential units and their related growths are not clear. Generally, areas covered by the residential units cannot be distinguished from commercial or institutional or other units by using remote sensing data. Digital classification of remote sensing imagery can identify the land cover classes (e.g., built-up) rather than land-use (e.g., residential, commercial). Further, number of households may also decrease with time (negative growth rate); for example, government of a congested city may purchase roadside housing units for a road widening project by demolishing the purchased housing units. In such cases, the preceding model will artificially show the development as dispersing. For the time span 1990–2000, Borough 5 has experienced a household growth rate of –13.42 %; however, the built-up growth rate was 12.57 %. If these values are combined, they show a compactness of –25.98 that may be misleading.

4.3 Results From Analysis: Based on Natural Boundary

This section is aimed to document and discuss the results obtained from the analysis based on the natural extents of the city.

4.3.1 Area Index and Shape Index

Table 4.4 shows the results obtained by applying area index. This index can provide us with the information of relative porosity—whether it is increasing or decreasing with change of time. The results show that the city is becoming more porous. However, as explained in Sect. 3.14.1 (Chap. 3), this index cannot say whether the city is sprawled or not, in black-and-white characterization.

Table 4.5 shows the results obtained from shape index for each date. This finding reveals that the complexity of shape was increasing during 1980–1990, then it was decreasing in the next decade, and again it was increasing during 2000–2011. Unfortunately, this finding contradicts with the finding from area index. As mentioned in Chap. 2, this problem in metric analysis is common. What to conclude now? Should we consider a few more metrics? How many metrics can be considered to be sufficient to conclude? What to do if they also contradict one

Table 4.4 Area of city-extent and built-up area within the city extent, percentage of built-up, and area index

Year	Area of city-extent (in ha)	Built-up area (in ha)	Percentage of built-up	Area index (area of city-extent/ built-up area)
1980	13002.92	9744.78	74.94	1.33
1990	18658.05	13830.30	74.13	1.35
2000	28813.03	21083.79	73.17	1.37
2010	36668.1	26258.30	71.61	1.40

Table 4.5 Perimeter, area, and shape index

Year	Perimeter (km)	Area of city extent (km^2)	Shape index (perimeter/area)
1980	390.43	130.03	3.00
1990	567.23	186.58	3.04
2000	663.63	288.13	2.30
2010	924.41	366.68	2.52

Table 4.6 Trend of urban growth during 1980–2010

Time span	City extent		Built-up area		City extent growth rate minus built-up growth rate (G_{CE}–G_B)
	Growth (in ha)	Rate of growth (G_{CE}) (in %)	Growth (in ha)	Rate of growth (G_B) (in %)	
1980–1990	5655.13	43.49	4085.52	41.93	1.56
1990–2000	10154.98	54.43	7253.49	52.45	1.98
2000–2010	7855.07	27.26	5174.51	24.54	2.72

another? These questions are yet to be answered in the research domain of spatial metrics.

The results of area index and shape index proved that they cannot be relied upon. Therefore, they have been rejected.

4.3.2 City Extent and Built-up Area

As proposed in Sect. 3.14.2 (Chap. 3), *city extent growth rate* minus *built-up growth rate* can provide a better insight into sprawl. Table 4.6 shows the results obtained from this index, in which positive values show sprawl and negative values show compactness. The degree of sprawl and compactness can also be identified by their magnitudes.

The results show that the city is becoming sprawled. As mentioned in Sect. 3.14.2 (Chap. 3), this index is an indirect measure of porosity. Therefore, it should support the results of area index. As it is evident in this research, they do support the results of each other. However, this index is better than the area index because it can identify the sprawl in black-and-white characterization.

Further, this index is also useful to analyze the growth rates of city extent and built-up areas within the city extents. Therefore, it can be used for the analysis of both urban growth and sprawl. As observed in Table 4.6, the growth rates in both the instances were increasing during 1980–2000. However, in the time span 2000–2010, they have registered approximate 50 % growth rates than that of the previous decade. Interestingly, the city became more porous despite low growth rates in city extent and built-up area. This clearly proves that the urban growth and sprawl should be measured with different scales. A low built-up growth rate does not always guarantee a compact development.

However, as mentioned in Sect. 3.14.2 (Chap. 3), this index considers the entire city in the analysis. It does not consider the variations of the built-up areas in the different parts of the city.

4.3.3 Shannon's Entropy and Built-up

The built-up areas within each zone (based on natural boundary, please refer Sect. 3.10, Chap. 3) and for each date have been calculated as presented in Table 4.7. If we calculate the entropy from Table 4.7, we can get the entropy values for the four temporal dates as shown in Table 4.8. The results show that the city of Kolkata was always sprawled, because the entropy values are always higher than the half-way mark of $\log_e(n)$. Relative entropy values are also always higher than the threshold 0.50. This finding supports the results of Sect. 4.3.2.

This model is robust, because it can identify the sprawl as a pattern for each date in black-and-white characterization. The threshold, that can determine whether the city is sprawled or compact, can also be determined mathematically. The change in entropy, between time t_1 and t_2, can be used to identify whether land development is becoming more dispersed (sprawled). Table 4.9 shows the values of change in entropy (ΔH_n) and change in relative entropy ($\Delta H_n^{'}$). As one can see, both are decreasing with time. Therefore, it can be said that the degree of sprawling rate is decreasing with time. However, decrease in sprawling rate should not be interpreted as 'compact'. If the difference in entropy (ΔH_n) or relative entropy ($\Delta H_n^{'}$) value for a specific time span becomes negative, then only it can be said that the city is becoming compact. The word 'compact' in this sense is not relative; rather, it is an absolute characterization of urban growth.

Relative entropy is free from scale effect of modifiable areal unit problem (MAUP). However, to mitigate the zone effect of MAUP, entropy decomposition theorem (please refer Sect. 3.14.3) has been applied. The notations and

Table 4.7 Built-up areas in different zones and different dates (in ha)

Year	Zone 1	Zone 2	Zone 3	Zone 4	Total
1980	9744.78	2059.72	1550.56	334.94	13690.00
1990	10148.41	3681.89	2536.13	771.07	17137.50
2000	10796.63	4825.07	5462.09	1719.81	22803.60
2010	11116.19	5001.51	5731.49	4409.11	26258.30

Table 4.8 Shannon's entropy for different temporal dates

Year	Entropy (H_n)	Relative entropy (H_n')	$\log_e(n)$	$\log_e(n)/2$
1980	0.86	0.62	1.39	0.69
1990	1.06	0.77		
2000	1.22	0.88		
2010	1.31	0.95		

Table 4.9 Differences in entropies between two successive dates under analysis

Time span	ΔH_n	$\Delta H_n'$
1980–1990	0.20	0.14
1990–2000	0.16	0.11
2000–2010	0.09	0.07

Table 4.10 Notation for entropy decompositions and the proportion of built-up areas

Year	Symbols		Values			
	Subscripts	j	1		2	
		i	1	2	3	4
		$i \in j$, n_j	$i \in 1$	n_1	$i \in 2$	n_2
1980	Proportions	P_j	0.8623		0.1377	
		$P_j(i)$	0.7118	0.1505	0.1133	0.0245
1990		P_j	0.8070		0.1930	
		$P_j(i)$	0.5922	0.2148	0.1480	0.0450
2000		P_j	0.6851		0.3149	
		$P_j(i)$	0.4735	0.2116	0.2395	0.0754
2010		P_j	0.6138		0.3862	
		$P_j(i)$	0.4233	0.1905	0.2183	0.1679

Table 4.11 Entropy decomposition analysis

Year	Relative entropies		
	H_n'	H_k'	$H_{n/k}'$
1980	0.62	0.58	0.67
1990	0.77	0.71	0.83
2000	0.88	0.90	0.86
2010	0.95	0.96	0.93

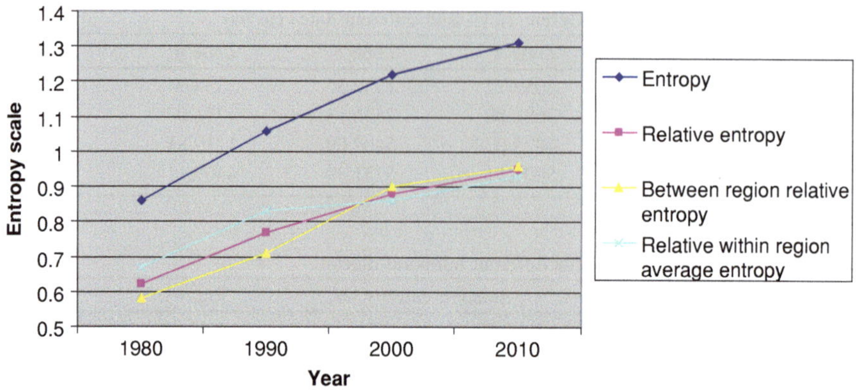

Fig. 4.6 Chart showing different entropies

proportions of built-up areas for each initial zone and delimited zone are presented in Table 4.10.

Table 4.11 shows the *relative entropy* (H_n'), *between region relative entropy* (H_k') and *relative within region average entropy* ($H_{n/k}'$) resulted from Eqs. 3.5, 3.12 , and 3.13 (Chap. 3). It is interesting to get different relative entropies, although the measures of entropy from Eqs. 3.4 and 3.9 (Chap. 3) are same. Important to note that the arithmetic mean of H_k' and $H_{n/k}'$ equals to H_n'.

Entropy decomposition theorem, which can eliminate the zoning problems associated with relative entropy, also reveals the same conclusion that the city of Kolkata is sprawled (since the relative entropy values are higher than 0.5). Both between-region and within-region relative entropies have shown dispersion in built-up distribution for all the image dates under analysis.

If we plot all the entropy values in a chart, as shown in Fig. 4.6, it shows that entropy values are continuously increasing. Therefore, it can safely be considered that the city of Kolkata is becoming more dispersed with the change of time.

4.3.4 Degree of Freedom in Urban Growth

Table 4.7 shows the observed built-up areas in different zones at different dates. From this table, the theoretical expected built-up areas can be calculated statistically by employing Eq. 3.14 (Chap. 3). Table 4.12 shows the expected built-up areas for different zones.

The expected and observed values can be used to determine the *degree of freedom* (Eq. 3.15, Chap. 3). It reveals the freedom or degree of deviation for the observed urban growth over the expected. For Table 4.7 (observed) and Table 4.12 (expected), the Chi-square test has resulted in Table 4.13 that reveals the degree of freedom as a pattern for each date under analysis.

Table 4.12 Expected built-up areas in different zones and different dates (in ha)

Year	Zone 1	Zone 2	Zone 3	Zone 4
1980	7163.96	2667.79	2618.46	1239.79
1990	8968.03	3339.62	3277.85	1552.00
2000	11933.09	4443.78	4361.59	2065.14
2010	13740.93	5117.00	5022.37	2378.00

Table 4.13 Degrees of freedom in urban growth for different dates

Year	Degree of freedom
1980	0.86
1990	1.06
2000	1.25
2010	1.31

Table 4.14 Differences in Chi-squares between two successive dates under analysis

Time span	Degree of freedom
1980–1990	0.20
1990–2000	0.19
2000–2010	0.05

The Chi-square (χ^2) has a lower limit of 0, when the observed value exactly equals the expected value. Higher magnitude shows higher degree of freedom. The measurement of the difference of Chi-square values between two dates can be used to indicate the magnitude of change in degree of freedom as presented in Table 4.14. This table shows that despite the increasing freedom its growth rate is declining.

This new approach of application can determine the freedom as a pattern as well as a process. Since this approach has used statistically derived expected growth that cannot be relied in all cases (as explained in Sect. 3.14.4), no comments have been made based on these metrics for the city of Kolkata.

4.3.5 Degree of Goodness in Urban Growth

The *degree of goodness* (the degree at which observed growth relates the planned growth and the magnitude of compactness in a single scale) has been calculated and results are furnished in Table 4.15. In this table, positive values indicate 'goodness', whereas negative values indicate 'badness'; a value of 0 is an indication of average condition.

The difference in goodness between two dates (Eq. 3.18, Chap. 3) has also been calculated to understand the goodness as a process. Table 4.16 shows the

Table 4.15 Degree of goodness in urban growth for different dates

Year	Degree of goodness
1980	0.62
1990	0.20
2000	−0.10
2010	−0.22

Table 4.16 Differences in goodness between two successive dates under analysis

Time span	Degree of goodness
1980–1990	−0.42
1990–2000	−0.30
2000–2010	−0.12

magnitude of change in degree of goodness between two successive dates under analysis. Table 4.15 reveals that the goodness is decreasing with time; however, Table 4.16 shows that the rate of decrease is declining. If the Eq. 3.18 results in a positive value, then it can be said the goodness is increasing.

However, since the preceding goodness analysis is based on the statistically derived expected growth, no comment on the city of Kolkata has been made by interpreting the presented results.

4.4 Comparison of Sprawl Analysis

Results obtained from several sprawl analysis techniques can be compared to draw the conclusion regarding the sprawl for the city of Kolkata. As presented in Table 4.17, most of the results show the city of Kolkata as sprawling. Entropy method, the most widely used and reliable technique, also shows the city is becoming sprawled rather than compact. Entropy decomposition theorem also proves that the *between-region* and *within-region* distributions of built-up are towards dispersion. Therefore, it can safely be concluded that the city of Kolkata is experiencing sprawled urban growth. The city was always sprawled; however, the growth rate of sprawl was decreasing. Important to mention that a declining growth rate in sprawl does not mean it is being compacted; because the compactness in this analysis has been derived in black-and-white absolute characterization, rather than relative. If the degree of compactness does not meet a mathematically derived threshold, it cannot be said the city is becoming compact. However, in relative sense, the city has shown a tendency of being less sprawled through time.

Many metrics that were in the use of earlier researches have not been considered in this analysis because of either their dependency on the spatial resolution of image data or of their redundancy in capturing the properties of urban growth/

Table 4.17 Summary of results from sprawl analysis

Index	Result
Analysis based on administrative boundary	
Difference between proportion of built-up area and proportion of population	Boroughs are in balanced condition
Difference between growth rates of population and built-up area	Different boroughs have different levels of compactness. KMC is sprawling
Difference between household growth rate and built-up growth rate	Different boroughs have different levels of compactness. KMC is sprawling
Analysis based on natural boundary	
Area index	Kolkata is sprawling
Shape index	Variable
Difference between city extent growth rate and built-up growth rate	Kolkata is sprawling
Entropy	Kolkata is sprawling
Relative entropy	Kolkata is sprawling
Between region relative entropy	Kolkata is sprawling
Relative within region average entropy	Kolkata is sprawling

sprawl. The metrics and scales, in this research, have been chosen carefully to avoid the stated limitations.

It has been found that the entropy method is the best metric for the analysis of urban sprawl. It can characterize the sprawl in black-and-white as a pattern as well as process. Although the entropy method is suffered from MAUP problems; these problems, however, can be mitigated. Relative entropy can overcome the scale effect of MAUP. The zoning effect can also be eliminated by the application of entropy decomposition theorem.

Table 4.17 also proves that traditional spatial metrics are inferior in capturing the sprawl phenomenon, as it was anticipated in the review of the literature. The shape index provides different results than that from other indices. Therefore, reliance on this metric is not justified.

4.5 Degree of Freedom and Goodness

The research has shown how to calculate the degree of freedom in urban growth in a different way than the earlier researches. This modified approach can determine the freedom as a pattern as well as a process. Finally, a modified approach has also been demonstrated that can determine the goodness of urban growth both as a pattern as well as process. Since these approaches are useful for the cities that have predetermined expected built-up values, no comments have been made based on these metrics for the city of Kolkata. However, in future, if the city of Kolkata experiences such planning policies, these metrics would be applicable. The preceding methods may provide with a new direction towards the analysis of urban growth.

The measure of goodness considers the relative entropy value as an indicator of sprawl, because entropy method is the most widely used and reliable measure of urban sprawl among the other available metrics. The relative entropy is free from the MAUP problems, and it is the arithmetic average of between-region and within-region relative entropies. Therefore, the relative entropy is the best sprawl measurement metric that can be used in quantifying the goodness of urban growth.

One may comment that the demonstrated approach of determining the goodness has a major limitation: it does not take into account any policy variables for the past. In this research, the expected urban growths have been calculated by a statistical approach based on the past and present urban growth observations, since there were no such predetermined and preplanned expectations. In many of the cities, they have predetermined and preplanned expectations of urban growth. Therefore, if one considers the predetermined expected values, the quantification of freedom and goodness will automatically be influenced by the policy variables. It is worth mentioning that although in the industrialized countries they may have proper planning policies for their cities; however, the cities in developing countries lack such type of policies in most of the cases and they grow with all freedoms. Therefore, the demonstrated approach of predicting the expected growth may be adopted for the cities in developing countries. However, the reliability of these statistically predicted values is not always guaranteed.

Whether a higher degree of goodness is an indication of sustainable development or not (could be debated since it should be judged with the empirical evidences of sustainability measures, but there is no doubt that lower freedom and lower sprawl are the general expectations). Since, the degree of goodness is computed from the products of these two variables, it is a direct measure of the goodness for urban growth.

4.6 Boundary of the City

Although the quantifications of sprawl, freedom, and goodness have been demonstrated on the natural extent of the city, they are equally applicable on the administrative boundaries and their subdivisions. They can also be applied on a study area that has been subdivided into concentric circles or pie sections. These methodologies have been tested and reported during research in Bhatta (2009) and Bhatta et al. (2010).

It has been observed that the analytical methods that are based on administrative tracts have several limitations in capturing the sprawl phenomenon. These models are useful for the analysis of urban growth, e.g., growth rates of population and built-up. The main reason to consider the administrative boundary is to embrace the census (demographic and socioeconomic) data in the analysis. This approach also has limitations, such as nonavailability of census data or delimitation of administrative boundaries. It has been evidenced from this research; census data for 1981 were incomplete and this data for 2011 were absent. Therefore, it can be said that, in many instances, the analysis is not possible if one considers the jurisdictional boundaries.

Another reason for the preference of administrative boundary is the regulation of urban growth by the civic bodies. The planning policies and growth control policies are generally enforced by the city administrators. Therefore, the analysis in consideration of administrative boundary facilitates the administrators to regulate these policies within their jurisdictions. A boundless analysis cannot help them in achieving their planning goals. This problem can be mitigated by establishing a civic body to regulate the planning policies within a much greater area than the core city; as it has been done in case of Kolkata by establishing Kolkata Metropolitan Development Authority (KMDA). Planning and growth control policies should be regulated by this larger body rather than the bodies having smaller areas under control.

Therefore, the natural extent of the city is better to consider in the analysis of urban growth and sprawl rather than the administrative boundaries. However, properly designed urban growth boundaries may also be considered for the planned cities instead of the natural boundary of the city.

4.7 Application of the Results

The research has resulted in a detailed report of the urban growth status for the city of Kolkata. These data can be used by the local administrators and planners to understand the past and present in order to plan and prepare for the future. The research has directed attention towards the magnitude and pattern of change in urban growth and sprawl of Kolkata for the last three decades, which would be very helpful in terms of guiding future planning policies for the city.

The results obtained from the research can also be utilized to provide some guidance as to the scale of change likely to result if trends of the past decades continue. If, for instance, past and present trends in built-up growth continue for the next 30 years, what levels of urban growth will this city has to prepare for, can be figured out using the observed percentage of increase in built-up area.

Although there are many reasons to neglect the study of past and present urban growth patterns to prepare for the inevitable future growth of cities, ultimately this practice results in the absence of even minimal preparation for urban growth, on both the activist and the regulatory fronts. It is, no doubt, an inefficient, inequitable and unsustainable practice, imposing great economic and environmental costs on societies that can ill afford them. This research demonstrates several methods/models by using remote sensing/GIS techniques to study the past and understand the present. Accordingly the future planning can be made. Our understanding of the complexity of urban growth pattern for the past and present essentially provides us with some of the tools necessary to meet it in an efficient, equitable, and sustainable manner in the years to come.

It is hopped that the city planners, administrators, local governments, and environmentalists of Kolkata will be benefited from the results of this research to ground their studies, debates, and decisions concerning these issues in empirical facts. The modified or proposed models and methods applied in this research may also be adopted and/or tested by the administrators and planers in other cities.

4.8 Limitations of the Research

The measures employed in this research are primarily based on built-up areas in each borough or zone. However, land areas available for development in each zone are not same. Therefore, it would be better if the measures could be calculated on the basis of percentage of built-up area within a zone by excluding non-developable land. This percentage can be calculated by the following formula:

$$p_i = \frac{B_i}{A_i - D_i^{\mathrm{N}}} \times 100 \tag{4.1}$$

where, p = percentage of built-up area, B = built-up area within a zone, A = total area of the zone, D^{N} = nondevelopable land within the zone, and i = a specific zone.

However, area of nondevelopable land cannot be measured directly from the remote sensing data. An area may be nondevelopable owing to several reasons that include natural (river, rugged terrain), ownership (land owned by military), government policy (to preserve an open space, water body, or agriculture), legal disputes on land, and several others. Therefore, it may be difficult to get such multi-temporal datasets. However, if these data are available, one may proceed by considering the aforementioned equation to calculate the percentage of built-up areas; and obviously will get more reliable results.

The approach of this study may be criticized due to its simplified approach for the analysis of urban growth pattern. Several arguments can be made owing to nonconsideration of road network, distribution of commercial centers, terrain properties, transition among different land-uses, and many more that have been considered by the other researchers. But, one has to remember that, in developing countries, cities have grown with unplanned developmental initiatives. In many instances, these cities lack historical data of urban development and multi-temporal inventory of land use/land cover data. Therefore, many of the spatio-statistical models cannot be adopted for these cities. Furthermore, in several instances, the city administrators are not well conversant with the new tools or methods and modern technologies such as geospatial technology. Therefore, they demand simple analytical approaches that require minimal set of input data. The demand of simplicity arises from this point of view. The demonstrated approaches, in this research, would be very helpful in terms of their simplicity in deriving historical data from satellite imageries and also in terms of computation.

In some of the instances, different approaches to detect and quantify the urban growth and sprawl (e.g., area index and shape index) have resulted in opposite conclusions. What may be the reasons behind these opposite conclusions would be an important issue of analysis. The current research mainly identifies the limitations in individual metrics. It does not attempt to find the statistical or other properties that may result in opposite conclusions. Such finding may provide a better insight into the urban growth and sprawl for a city.

Although the demonstrated models can quantify the urban growth and sprawl; however, they cannot provide insights into reasons behind or the impacts of such growth. Therefore, to understand the causes or consequences, we need to develop several other models. The causes that force growth in urban areas and the causes that are responsible for undesirable pattern or process of urban growth are also essentially important for the analysis of urban growth. The consequences or the impacts of urban growth, whether ill or good, are also necessary to be understood and evaluated towards achieving a sustainable urban growth. The impacts of development present specific development patterns as undesirable, not the patterns themselves. Therefore, whether a pattern is good or bad should be analyzed from the perspective of its consequences. Causes are also similarly important to know the factors that are responsible to bring such a pattern. However, remote sensing data are inferior in the analysis of causes or consequences; for example, what may be the impact of a specific pattern of urban growth on human life or on other animals or on economic base of the city cannot be viewed by the remote sensing sensors. In some of the instances, inferences can only be made.

This research also fails to provide any acceptable approach to encounter the problems associated with urban growth and sprawl, or for how to restrict them, or to clearly demarcate the merits/demerits of imposing restrictions on urban growth for the city of Kolkata.

One of the objectives of this research was to modify existing models or to introduce new tools or techniques for the analysis of urban growth so that they can be applied universally. However, the models to measure the freedom and goodness do not fit unplanned cities (the ones that do not have any preplanned values of urban growth). Cities in developing countries are mainly unplanned; therefore, the scope of application is very limited for these models. From this observation it can be said that, without having any ancillary data, remote sensing data alone cannot serve all analytical purposes.

Finally, this type of analysis is misguided by all of the limitations of remote sensing imageries that form the basis of information. These limitations and their possible effects have been documented with a great detail in Bhatta (2010). An important consideration of remote sensing data is the spatial accuracy or spatial resolution. Data accuracy and resolution directly affect landscape heterogeneity. This issue is central to all remote sensing data analysis. Different sensors generate different types of images that are often to be considered in multi-temporal analysis that result in different images to be compared. Different image processing techniques (or algorithms) may also generate different results for the same image. Different classification schemes (level of classification) also generate different resultant maps. Further, the generation method of ancillary data that are often required for analytical or other purposes (e.g., validation of accuracy) may also vary in a wide spectrum. Use of these varying data generation methods is a common practice in urban analysis; because data generation methods and resulting data are not designed to be consistent with different datasets. However, the question arises regarding the validity of directly comparing spatial data derived from different data generation processes to determine the magnitude, location, and pattern of development.

Although classified satellite imagery, at a range of scales, has long been used in urban applications, classification of land covers within an urban landscape is a difficult task. Urban areas are highly heterogeneous; generally there are many landscape features present in a small transact. This results in virtually every pixel of even high spatial resolution imagery being a mixture of a vast range of different surfaces (mixed pixel problem). In addition, identifying accurately which pixel matches which area on the ground can represent a difficult problem, leading to registration errors. Error and uncertainty in remote sensing-based land cover maps also represent a major drawback to operational application. The classification capabilities of remote sensing data mainly depend on the spectral contrast between the classes of interest and the spectral resolution of the sensor. The lower spectral separability of classes to be determined means less accuracy in the classified land cover maps. Classification of remote sensing data to urban scenes is therefore fraught with difficulties, especially when attempting to segment the typically heterogeneous image structures.

However, despite several limitations, this study shows how to quantify the sprawl, freedom, and goodness of urban growth as a *process* as well as a *pattern*. The methods and models that have been demonstrated in this research are extremely less demanding in terms of data and computation; especially, if one considers the analysis in consideration of natural boundaries, remote sensing imageries are the only demanded data. It is hoped that the exercises of this research will ground the knowledge of interested administrators and politicians for their cities, in developing as well as developed countries.

References

Bhatta B (2009) Analysis of urban growth pattern using remote sensing and GIS: a case study of Kolkata, India. Int J Remote Sens 30(18):4733–4746

Bhatta B (2010) Analysis of urban growth and sprawl from remote sensing data. Springer, Heidelberg, p 172

Bhatta B, Saraswati S, Bandyopadhyay D (2010) Quantifying the degree-of-freedom, degree-of-sprawl, and degree-of-goodness of urban growth from remote sensing data. Appl Geogr 30(1):96–111

Census (1981) District Census Handbook, Series 23, Parts XIII A&B. Directorate of Census Operations, West Bengal

Census (1991) District Census Handbook, Series 26, Parts XII A&B. Directorate of Census Operations, West Bengal

Census (2001) Table 4, Table 12, Table 13, and Table B21. Directorate of Census Operations, West Bengal

Chapter 5
Conclusions

Abstract This chapter aims to draw the conclusions and to highlight the future scope of research and study. It draws conclusions not only on the urban growth and sprawl for the city of Kolkata, but also on the methods and metrics used or modified or proposed in the current research. Future research directions have also been made based on the limitations of the current research (documented in Chap. 4). This research may aid into help local authorities in general to establish and evaluate development goals in terms of establishing social and other infrastructure equipments considering the threats of urban sprawl in achieving sustainable development. Apart from the derived conclusions, the research also shows the potentials of remote sensing and GIS for the analysis of urban growth and sprawl in spatio-temporal scales.

Keywords Urban growth · Sprawl · Remote sensing · GIS · Kolkata · India · Metrics

5.1 Conclusions

Based on the results and discussion, the following conclusions can be made:

1. The rate of urban growth in the city of Kolkata is generally higher than the growth rate of population as well as growth rate of household. The analysis shows a clear indication of urban sprawl in the city of Kolkata. However, it has also been noticed that the degree of sprawling rate is declining with time. It is significant to note that Kolkata, being a metropolitan city of a developing country, is not becoming more compact with the declining rate of population growth. These observations suggest that there may be an urgent need of restricting the sprawl of the city in view of the environmental and ecological sustainability (because sprawl, in general, threatens the environmental and ecological sustainability).

B. Bhatta, *Urban Growth Analysis and Remote Sensing*, SpringerBriefs in Geography,
DOI: 10.1007/978-94-007-4698-5_5,

2. In this research, several existing metrics, for the analysis of urban growth and sprawl, have been reviewed and tested. It has been found that most of the existing metrics are highly dependent on the spatial resolution of the image data, and often they measure redundant properties. Most of these methods demand huge data and computation and even, there may be substantial contradict in conclusion among these methods. Further, many of the existing metrics cannot identify compactness as a pattern as well as process in black-and-white absolute measures. Entropy method has been proved to be the most stringent measurement tool among the tested sprawl measurement techniques. However, it seems that some other methods should also be considered to justify the acceptance of the entropy method. In this regard, black-and-white characterization of sprawl may be preferred rather than many relative measures.
3. In the present study, modifications of several sprawl measurement metrics have been developed (as described in Chap. 3). It has been found that results obtained from these modified models are in tune and supplementing each other. Therefore, their acceptance is justified, and they may be tested on other cities. Pearson's Chi-square statistic model to determine the degree of freedom has also been used in a modified approach that can analyze both the process as well as pattern of urban growth.
4. The research has attempted to measure the goodness in urban growth in a new approach. However, it seems that it may be more suitable for the cities having well-defined and preplanned expected urban growth.
5. The research has tested several feasible models for the measurement of sprawl in the city of Kolkata (refer Table 4.17, Chap. 4). Most of the demonstrated models provided similar results except the model based on shape index. However, the analysis shows that 'difference between household growth rate and built-up growth rate' seems to be more stringent compared to other analytical techniques based on the administrative boundary. Considering the natural boundary, the existing 'entropy method' and the developed 'difference between city-extent growth rate and built-up growth rate' seem to be more accurate. These models, based on remote sensing data, have proved to be useful for the identification of urban growth pattern and their general tendencies. These models are less demanding in terms of data, computation, and expertise. They are simple and empirically devised, not constrained by the straitjacket of rigid theory devices, and they can identify the sprawl in black-and-white characterization. Further, the study has found that the natural boundary of the city is more preferred than the administrative boundary for the analysis of urban growth and sprawl.
6. A new zoning concept (as explained in Sect. 3.10, Chap. 3), based on multi-temporal natural boundary, has also been proposed in this study. This zoning concept is completely based on the remote sensing data, rather than administrative boundary or any other hypothetical geometry. It considers the boundary of the city as a dynamic phenomenon; therefore, it is more rational and reliable compared to the other available zoning techniques.

7. The results and analysis in this study have attempted to draw the attention of the magnitude and pattern of urban growth of Kolkata for the last three decades. The results obtained from this research may also be utilized to forecast the future characteristics in terms of urban growth and sprawl assuming the same trends of the past decades.

In summary, the study opened with an observation about the important roles of analytical models for urban growth pattern and process from remote sensing data, proceeded to use some standard classical and some new models for analytical purposes. The theory and models of urban spatial growth analysis, which are supported by the findings of this research, may prove useful in analyzing the urban growth of a city for the past and present by using remote sensing data. However, there still remains some important issues (addressed in the next section) for which analytical support has to be made available. Further analytical research may shed some light on these issues in the near future.

5.2 Scope of Future Research

The derived results from this study should interest some researchers to derive suitable models toward establishing the relationships between the reasons of urban growth/sprawl and the results of this research. The consequences of such growth in Kolkata may also form the basis of further research. Study of socioeconomic and environmental impacts of urban growth and sprawl for the city of Kolkata can ground better knowledge for the understanding of the problems associated with urbanization in order to plan for a sustainable future. Research has already been initiated, during this study, to model the relationships between the urban growth and causes/consequences. One such model (that relates number of working persons, developable lands within a zone and built-up growth) has been reported in Bhatta (2009). The model shows how the built-up area increases with the increase in number of working persons in consideration of available land for new construction.

Another hypothesis might serve as the basis for future research: higher degree of goodness is an indication of sustainable urban growth. Sustainability is a process which tells of a development of all aspects of human life affecting sustenance. It means resolving the conflict between the various competing goals, and involves the simultaneous pursuit of *economic prosperity*, *environmental quality* and *social equity* (Bhatta 2010). Therefore, it is obvious that degree of goodness cannot be a direct measure of sustainable development. Rather, the degree of goodness may become one of the indicators of sustainable development. However, before drawing such conclusion, one has to reliably establish the correlations among them.

The model for measuring the goodness and the approach for measuring the freedom could not been empirically tested and justified in this research owing to the lack of preplanned expected values of urban growth for the city of Kolkata. These models may be applied on cities having these predetermined values to justify their acceptance or rejection. It is expected that future researches will come out with the justification of reliance on these models for the cities having predetermined expected values.

The key issue facing decision makers at the local, national and international levels, is not whether the urban growth or sprawl will take place, but rather what is likely to be the scale of urban growth or sprawl and what needs to be done to adequately prepare for it. The current research presents only the analysis of urban growth. What the measures should be taken, in order to mitigate the problems associated with undesirable urban growth will essentially be an important issue of future research.

References

Bhatta B (2009) Analysis of urban growth pattern using remote sensing and GIS : a case study of Kolkata, India. Int J Remote Sens 30(18):4733–4746

Bhatta B (2010) Analysis of urban growth and sprawl from remote sensing data. Springer, Heidelberg, p 172

Appendix A
The City of Kolkata

This appendix provides a brief description about the city of Kolkata—the study area of the research documented in this book.

A.1 Introduction

The city of *Kolkata* (formerly *Calcutta*) is more than 300 years old and it served as the capital of India during the British governance until 1911. Kolkata is the capital of the Indian state of West Bengal; and is the main business, commercial, and financial hub of eastern India and the north-eastern states. It is located in the eastern India at 22° 33′N 88° 20′E on the east bank of *River Hooghly* (Ganges Delta) (Fig. A.1) at an elevation ranging from 1.5 to 9 m (SRTM image, NASA, Feb 2000).

A.2 Administrative Structure

The civic administration of Kolkata is executed by several government agencies, and consists of overlapping structural divisions. At least five administrative definitions of the city are available:

1. *Kolkata Central Business District*: hosts the core central part of Kolkata and contains 24 wards of the municipal corporation.
2. *Kolkata District*: contains the center part of the city of Kolkata. It is the jurisdiction of the Kolkata Collector.
3. *Kolkata Police Area*: the jurisdiction of the Kolkata Police covers the KMS and some adjacent areas as well.[1]

[1] The service area of Kolkata Police was 105 km^2 as of 31st Aug 2011. The area has been extended from 1st Sep 2011 to cover the entire KMC and some adjacent areas.

B. Bhatta, *Urban Growth Analysis and Remote Sensing*, SpringerBriefs in Geography, DOI: 10.1007/978-94-007-4698-5,

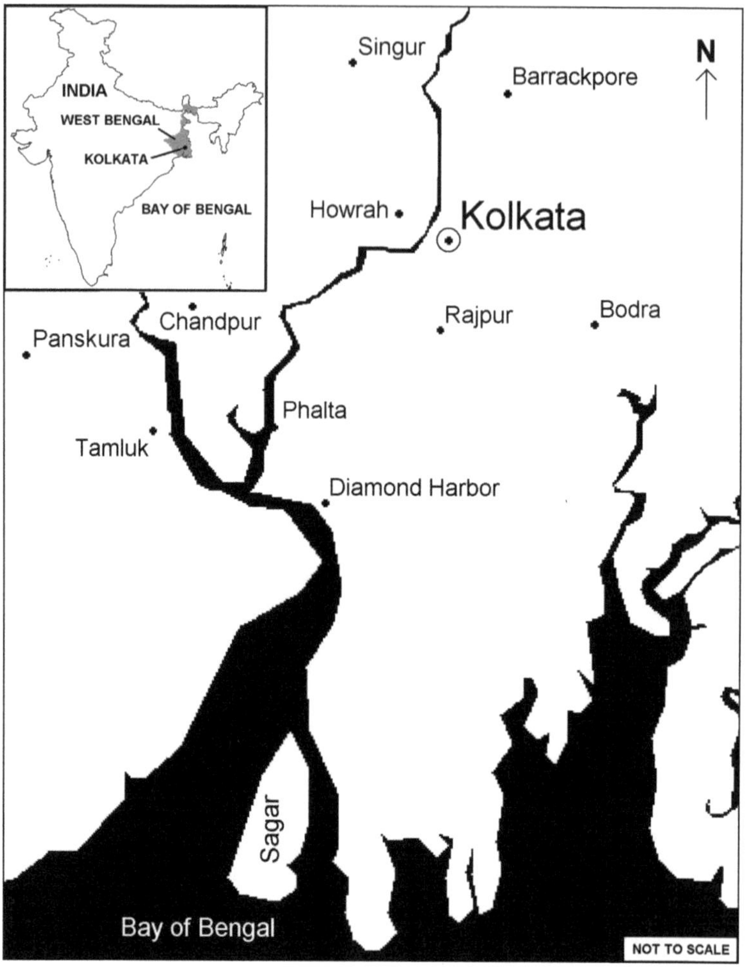

Fig. A.1 Location of Kolkata

4. *Kolkata Municipal Corporation (KMC) Area*: covers the most of Kolkata Police area (as shown in Fig. 3.1 in Chap. 3); served and administrated by KMC.
5. *Kolkata Metropolitan Area (KMA)*: the urban agglomeration of the city of Kolkata. Kolkata Metropolitan Development Authority (KMDA) is the statutory planning and development authority for the KMA functioning under the administrative control of Urban Development Department of Government of West Bengal. The KMDA area consists of 3 municipal corporations (*Kolkata*, *Howrah*, and *Chandannagr*), 38 independent municipalities, 77 towns, and 445 villages (shown in Fig. A.2).

Table A.1 Boroughs and their associated wards of KMC (as on 23 July 2010)

Borough No.	Associated wards	Borough No.	Associated wards
1	1–9	9	74–80, 82, 83, 88
2	10–12, 15–20	10	81, 89, 91–100
3	13, 14, 29–35	11	101, 102, 110–114
4	21–28, 38, 39	12	103–109
5	36, 37, 40–45, 48–50	13	115–123
6	46, 47, 51–55, 60–62	14	124–132
7	56–59, 63, 64, 66, 67	15	133–141
8	65, 68–73, 84–87, 90		

On 24 July 2010, the House of Councilors of KMC has passed a resolution for a delimitation plan of boroughs. According to the new plan, wards 71 and 73 of borough 8 will be included in borough 9 and these two wards will be replaced by wards 82 and 83, previously falling under borough 9. Similarly, wards 101 and 102 of borough 11 will be included in borough 12 and these wards will be replaced by wards 103 and 104 of borough 12.

The word 'city', in relation to Kolkata, is not an official term; however, when used, it normally refers to the KMA. It may seem paradoxical that the area of a district is a subset of one urban area; normally a district contains several urban areas as well as rural areas within it.

A.3 Kolkata Municipal Corporation (KMC) Area

The KMC (formerly *Calcutta Municipal Corporation* or CMC) is responsible for the civic infrastructure and administration of the core city. The KMC area is divided into 141 administrative wards (Fig. 3.1) that are grouped into 15 boroughs (refer Fig. 3.2 in Chap. 3 and Table A.1) (KMC 2009). Boroughs 11–15 spread across the southern, south-eastern, and western fringes of the city are commonly called 'added area'. In early 1985, the Government of West Bengal included South Suburban Municipality, Gardenreach Municipality, and Jadavpur into KMC. After the delimitation in 1985, these areas constitute boroughs 11–15.

The KMC has a jurisdictional area of 187.33 km^2 (KMC 2009). This area is spread linearly along the east bank of the River Hooghly in a north-south direction. The east-to-west dimension of KMC is stretching from the River Hooghly in the west to roughly the Eastern Metropolitan Bypass in the east. Much of the city (mainly in the eastern part) was originally a vast wetland, reclaimed over the decades to accommodate the city's burgeoning population.

A.4 Kolkata Metropolitan Area

KMA is the urban agglomeration of the city of Kolkata. It consists of many urban municipalities as well as rural areas. Current extent of KMA is 1851.41 km^2

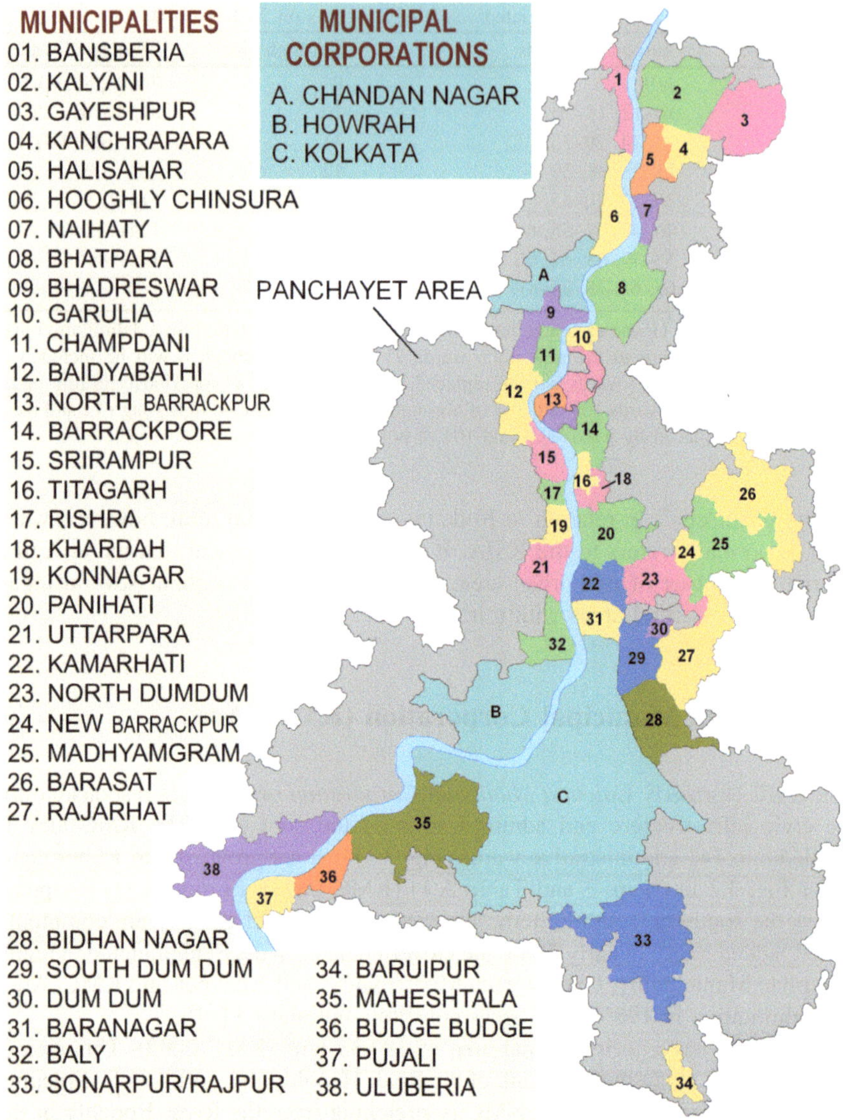

Fig. A.2 Kolkata Metropolitan Area (Courtesy: KMDA)

(Fig. A.2, Table A.2). It is the largest urban agglomeration in eastern India, and second largest in India with a population of 13.2 million as per 2001 census. This area contains more than 50 % of the total population of West Bengal state. The core areas of KMA such as Kolkata and Howrah Municipal Corporation areas show one of the lowest population growth rates during 1991–2001 compared to the growth rates of 1981–1991. On the other hand, the peripheral areas show a reversal

Table A.2 Composition of KMA 2001

Categories of area	Numbers	Area in km^2
Municipal corporations	3	271.31
Municipalities	38	615.49
Non-municipal urban/census towns	77	200.10
Outgrowths	16	18.19
Rural area	445	746.32
Total KMA		1851.41

Source Website of Jawaharlal Nehru National Urban Renewal Mission, Ministry of Urban Development, Govt. of India (http://jnnurm.nic.in/nurmudweb/toolkit/KolkataCdp/CH-I1-17.pdf).

in the growth pattern of populations in the decade 1991–2001 compared to the growth rates of core areas. It may be seen from census data that the peripheral areas registered three times growth in 1991–2001 compared to 1981–1991. The population of KMA is projected at 17 million in 2011, 20 million in 2021, and 21.1 million in 2025 (KMDA 2009).

The KMA is governed by many local governments in different administrative zones. However, for the overall management of the development of this area, a statutory authority (KMDA; formerly CMDA) was established in the year 1970. KMDA is engaged in carrying out different levels of planning-exercises right from the perspective plan to formulation of investment programs in the KMA.

A.5 Urban Structure of Kolkata

The most visible element of the infrastructure, i.e. the street network is quite inadequate in Kolkata (Puchera et al. 2005). The road space (matched with population density) in the city is only 6 %, compared to 23 % in Delhi and 17 % in Mumbai, creating major traffic problems (The Hindu Business Line 2004). In comparison, cities with more than 2,50,000 populations in the United States have, on an average, 28.9 % of the city surface devoted to streets (Banerjee 2005).

The pressure of population in Kolkata has been felt on urban housing. The total occupied housing units of KMA in 1981 was around 1.7 million, this had increased to 2.2 million in 1991, and 2.75 million in 2001; for a city of over 5.7 million people in 1981, 8.5 million people in 1991, and 13.21 million in 2001; thereby reflecting an acute shorting of housing facilities. The total number of household of KMC in 2001 was 9,29,586, which is clearly inadequate for 45,72,876 people (Table A.3).

It has been noticed that in some of the wards the number of households and populations have been decreased (please refer Appendix B), especially in the northern and central Kolkata. These areas are the oldest urban areas of the city. Recently many residential buildings in these areas are being converted into business or commercial buildings. Negative population or household growth rates may be the consequence of that conversion.

Table A.3 Number of households and population of Kolkata (in million)

	KMA[a]		KMC[b]	
Year	Households	Population	Households	Population
1981	1.70	5.70	0.76	4.1
1991	2.20	8.50	0.85	4.4
2001	2.75	13.21	0.92	4.6

Source Census of India

[a] Approximate values

[b] Household and population figures of KMC for 1981 includes the areas of South Suburban, Gardenreach, and Jadavpur

The most pressing of all problems associated with the population growth and urban sprawl had been the appearance of slums and squatters. Their appearance had started right from the British Colonial days, and had been more and more aggravated by acute geographic, social, and ethnic disparities in the mega city. As of 2001 census, one-third of the population lives in deprivation from basic amenities in highly congested and degraded slums. These slums occupy large tracts of land in the metropolitan core. Simultaneously, uncontrolled urban sprawl and unmet housing requirements endangers the ecologically sensitive wetlands bordering the city. One of the most blighted areas of the urban agglomeration is the twin city of Howrah, which itself has million plus populations, and is historically neglected area. Over half of the population here lives in slums (Banerjee 2005).

Since, several parallel and hierarchical governments and/or development agencies work together within this area; the city has experienced uncoordinated and patchy urban development. The city also lacks any master plan for the development in consideration of entire KMA. Till date no boundary has been delineated or no other restrictions have been imposed to stem its uncontrolled growth.

Appendix B
Census Data

Table B.1 Census data for municipal wards of Kolkata Municipal Corporation (KMC) (***Courtesy*** Directorate of Census, India; Census 1981, 1991, 2001)

Ward No.	Number of households in 1981	Number of households in 1991	Number of households in 2001	Population in 1981	Population in 1991	Population in 2001
1	10,302	10,571	10,038	5,0973	52,499	49,018
2	8,784	9,198	10,486	44,938	49,892	47,327
3	10,222	9,844	10,799	55,680	54,147	53,299
4	8,149	8,630	8,642	43,523	41,505	40,121
5	5,044	5,149	5,322	28,397	27,581	26,117
6	8,957	8,181	8,750	42,626	40,273	48,096
7	2,201	3,888	4,479	22,667	19,809	20,226
8	4,686	4,279	4,366	26,203	23,306	21,071
9	4,355	4,416	4,265	25,549	23,687	20,374
10	5,300	5,660	5,872	30,596	29,679	33,807
11	4,860	5,327	5,260	29,063	29,105	26,190
12	4,623	5,437	5,701	26,481	28,952	28,912
13	5,603	5,543	6,684	28,840	26,912	31,118
14	10,150	10,834	10,986	52,089	53,855	49,698
15	5,135	5,442	5,803	30,974	32,296	26,709
16	4,170	4,203	4,263	22,690	23,172	26,665
17	5,078	5,394	5,062	28,951	29,424	24,016
18	5,627	5,563	5,594	25,131	24,556	25,508
19	4,592	4,960	5,488	23,747	25,509	24,479
20	4,135	4,119	4,271	22,634	23,836	20,481
21	4,809	4,847	4,413	23,900	24,980	24,629
22	2,488	3,303	3,218	21,941	19,864	20,569
23	5,026	4,232	3,797	29,929	26,530	21,775
24	3,820	3,523	3,558	20,954	19,713	19,520
25	7,256	7,448	7,654	42,258	43,149	39,160
26	6,063	6,473	6,023	30,849	34,145	31,400

(continued)

DOI: 10.1007/978-94-007-4698-5,

(continued)

Ward No.	Number of households in 1981	Number of households in 1991	Number of households in 2001	Population in 1981	Population in 1991	Population in 2001
27	3,873	4,188	4,831	22,479	23,543	23,089
28	7,206	7,030	7,286	43,694	44,507	41,775
29	7,291	7,716	7,446	42,491	47,598	46,887
30	5,432	5,823	5,932	29,574	30,422	28,278
31	7,275	7,379	7,606	35,449	35,992	36,456
32	7,648	8,717	9,996	43,295	41,477	46,081
33	7,553	8,026	9428	43718	42108	44230
34	6,902	7,333	6,266	38,354	34,948	27,808
35	7,349	7,385	6,746	38,960	35,633	31,320
36	5,441	6,505	3,871	32,438	28,734	22,851
37	4,503	4,600	4,275	27,822	24,187	24,004
38	6,498	6,115	5,449	38,126	35,202	28,083
39	3,730	3,502	4,158	24,000	25,558	28,255
40	5,711	5,698	4,948	31,895	28,403	24,384
41	4,086	4,175	3,814	24,745	29,174	25,486
42	5,184	4,816	4,169	27,150	28,862	26,077
43	3,853	3,827	4,083	24,654	24,446	29,647
44	4,904	5,255	5,061	33,623	33,773	33,652
45	2,483	2,996	2,843	12,357	16,025	15,360
46	4,000	3,679	3,849	21,298	20,986	22,959
47	3,733	4,293	3,783	20,087	23,653	21,218
48	6,255	5,767	5,159	27,111	25,472	21,856
49	3,544	4,264	3,601	23,121	19,945	20,671
50	4,779	4,861	3,861	26,219	23,367	18,365
51	4,013	3,933	3,666	21,882	19,519	16,272
52	3,541	3,920	4,185	21,536	22,493	24,672
53	5,500	5,232	5,509	31,342	29,077	28,910
54	5,497	6,132	6,612	35,885	39,077	40,299
55	5,054	6,208	6,946	31,109	31,235	35,955
56	8,067	8,212	7,901	42,655	43,948	43,304
57	8,558	10,124	9,236	40,301	48,951	45,206
58	9,097	15,702	16,257	44,917	78,565	86,487
59	9,630	12,524	12,687	48,669	63,623	66,690
60	6,821	6,560	7,256	42,231	41,040	42,585
61	6,734	6,515	5,857	37,626	35,211	34,128
62	5,626	6,252	7,484	33,678	36,282	46,640
63	7,213	6,456	6,307	36,155	32,554	32,123
64	5,307	5,914	4,627	33,868	31,041	26,891
65	9,179	12,096	13,292	59,916	70,846	80,098
66	7,431	11,308	13,420	39,173	62,621	70,179
67	6,829	10,302	11,662	35,218	49,679	54,380
68	5,681	5,955	5,702	28,176	26,457	24,181
69	9,477	9,700	8,696	5,2134	4,8938	43,358

(continued)

(continued)

Ward No.	Number of households in 1981	Number of households in 1991	Number of households in 2001	Population in 1981	Population in 1991	Population in 2001
70	5,158	6,154	6,575	33,646	30,893	31,774
71	6,827	6,915	7,271	37,229	35,656	33,199
72	5,211	5,333	5,159	30,693	27,094	24,487
73	5,576	5,048	5,456	28,018	24,876	24,416
74	6,563	7,247	7,123	38,733	35,518	37,119
75	5,566	5,170	4,806	26,441	22,996	24,392
76	6,517	6,185	5,106	31,897	28,962	24,248
77	8,438	8,111	7,802	49,420	45,987	44,071
78	8,976	11,226	10,668	55,675	58,444	58,930
79	12,209	8,982	8,676	46,144	42,775	42,229
80	11,817	6,962	7,837	40,481	36,305	38,587
81	7,376	10,028	9,897	43,817	49,285	47,258
82	8,798	9,322	9,469	47,236	43,790	43,347
83	6,603	6,134	4,952	35,232	31,739	24,381
84	4,535	4,694	4,769	27,303	23,710	23,400
85	6,558	6,970	6,762	33,200	32,477	31,231
86	5,607	5,498	5,727	26,078	25,689	25,148
87	4,148	3,846	3,121	21,497	18,534	13,324
88	6,055	6,119	6,002	30,902	28,378	27,050
89	4,923	5,925	5,836	24,955	29,754	26,781
90	4,419	4,870	5,264	21,555	21,248	22,145
91	6,032	6,906	8,389	32,388	34,439	36,453
92	3,937	7,890	8,645	31,758	38,127	35,916
93	9,573	11,727	12,386	46,529	54,493	56,029
94	4,069	4,787	6,295	21,109	22,702	29,570
95	3,870	6,229	7,123	21,459	28,778	28,500
96	5,126	6,215	7,207	27,851	29,443	28,990
97	4,892	7,456	8,771	24,800	35,944	37,404
98	4,482	6,143	7,297	24,960	29,520	30,514
99	3,369	3,840	4,448	19,221	19,580	19,855
100	5,088	5,892	7,359	27,207	27,850	29,665
101	–	6,563	8,946	–	30,896	37,634
102	–	4,346	4,854	–	21,240	20,646
103	–	4,708	6,324	–	21,417	25,528
104	–	5,433	7,374	–	25,527	29,459
105	–	3,903	4,881	–	18,994	20,980
106	–	4,613	7,124	–	22,729	30,500
107	–	5,880	9,335	–	27,616	39,730
108	–	3,717	9,334	–	18,326	38,338
109	–	4,693	8,756	–	23,271	37,610
110	–	4,523	5,798	–	20,776	22,765
111	–	4,765	7,,556	–	23,003	32,149
112	–	4,513	5,907	–	21,547	25,497

(continued)

(continued)

Ward No.	Number of households in 1981	Number of households in 1991	Number of households in 2001	Population in 1981	Population in 1991	Population in 2001
113	–	4,714	7,316	–	22,674	30,933
114	–	5,053	7,018	–	24,875	31,416
115	–	5,584	7,074	–	27,108	30,616
116	–	5,257	6,461	–	26,274	28,473
117	–	4,980	5,081	–	24,426	25,276
118	–	5,056	5,938	–	23,935	25,922
119	–	3,970	4,074	–	18,863	19,410
120	–	4,460	5,155	–	21,162	21,900
121	–	5,951	6,156	–	27,829	29,970
122	–	5,576	7,996	–	27,561	34,957
123	–	4,739	7,215	–	22,732	30,569
124	–	5,911	7,560	–	26,556	34,295
125	–	6,285	9,984	–	30,016	42,245
126	–	4,835	6,465	–	23,301	27,234
127	–	5,716	8,706	–	27,672	36,584
128	–	5,385	7,520	–	25,412	31,864
129	–	6,610	9,087	–	31,857	38,967
130	–	5,077	5,980	–	23,314	26,630
131	–	5,753	7,301	–	25,522	30,136
132	–	5,458	6,771	–	24,028	28,089
133	–	4,588	4,360	–	22,975	23,881
134	–	5,766	5,869	–	31,462	36,687
135	–	4,710	5,019	–	28,822	31,743
136	–	3,726	4,029	–	19,837	21,810
137	–	4,024	2,974	–	18,592	20,036
138	–	4,170	4,340	–	26,051	34,724
139	–	4,605	6,210	–	33,114	45,006
140	–	2,685	3,935	–	21,224	29,636
141	–	3,753	5,048	–	24,106	31,183
South Suburban Municipality	77,465	–	–	3,94,916	–	–
Gardenreach Municipality	32,921	–	–	1,91,107	–	–
Jadavpur Municipality	46,483	–	–	2,51,968	–	–
KMC	7,59,110	8,53,337	9,29,586	41,26,139	43,85,176	45,72,876

Table B.2 Census data for municipal boroughs of KMC (***Courtesy*** Directorate of Census, India; Census 1981, 1991, 2001)

Borough No.	Number of households in 1981	Number of households in 1991	Number of households in 2001	Population in 1981	Population in 1991	Population in 2001
1	62,700	64,156	67,147	3,40,556	3,32,699	3,25,649
2	43,520	46,105	47,314	2,40,267	2,46,529	2,36,767
3	65,203	68,756	71,090	3,52,770	3,48,945	3,41,876
4	50,769	50,661	50,387	2,98,130	2,97,191	2,78,255
5	50,743	52,764	45,685	2,91,135	2,82,388	2,62,353
6	50,519	52,724	55,147	2,96,671	2,98,573	3,13,638
7	71,311	92,638	95,389	3,80,872	4,81,828	5,05,358
8	63,197	64,983	64,502	3,39,529	3,15,572	2,96,663
9	81,542	75,458	72,441	4,02,161	3,74,894	3,64,354
10	62,737	83,038	93,653	3,46,054	3,99,915	4,06,935
11	–	34,477	47,395	–	1,65,011	2,01,040
12	–	32,947	53,128	–	1,57,880	2,22,145
13	–	45,573	55,150	–	2,19,890	2,47,093
14	–	51,030	69,374	–	2,37,678	2,96,044
15	–	38,027	41,784	–	2,26,183	2,74,706
South Suburban Municipality	77,465	–	–	3,94,916	–	–
Gardenreach Municipality	32,921	–	–	1,91,107	–	–
Jadavpur Municipality	46,483	–	–	2,51,968	–	–
KMC	7,59,110	8,53,337	9,29,586	41,26,139	43,85,176	45,72,876

Note The boundary of KMC has been delimited in the year 1985. Earlier the KMC was constituted with 100 wards (10 boroughs). Later the South Suburban, Gardenreach, and Jadavpur municipalities have been included within the KMC area. The census data for the year 1981 are not available individually for the wards 101–141 (boroughs 11–15). However, if one includes the data of aforementioned three municipalities, the summed data can represent the current boundary of KMC

Appendix C
Built-up Data

Table C.1 Built-up data for boroughs of Kolkata Municipal Corporation (in hectare)

Boundary	Total area	Built-up area in 1980	Built-up area in 1990	Built-up area in 2000	Built-up area in 2010
Borough 1	947.49	632.70	694.86	792.67	808.11
Borough 2	326.84	279.64	301.43	316.11	317.65
Borough 3	938.86	573.95	656.72	756.85	765.90
Borough 4	335.63	293.59	311.90	324.57	324.65
Borough 5	530.61	424.02	433.01	487.43	487.55
Borough 6	556.30	456.59	460.06	508.71	509.84
Borough 7	2,853.84	912.47	1,007.44	1,284.49	1,441.42
Borough 8	990.06	714.33	735.89	838.07	844.77
Borough 9	1,938.15	1,122.72	1,220.81	1,439.71	1,469.72
Borough 10	1,582.79	659.96	911.10	1,135.68	1,263.24
Borough 11	1,120.63	57.92	167.42	489.79	652.58
Borough 12	2,231.24	122.88	242.87	659.62	1,051.25
Borough 13	1,542.98	361.25	533.48	759.95	916.61
Borough 14	1,985.84	264.78	517.57	788.94	1,013.51
Borough 15	864.08	347.34	381.11	508.22	522.63
KMC	18,745.34	7,224.14	8,575.67	11,090.81	12,389.43

Table C.2 Built-up data for the natural city-extents (in hectare)

City-extent of year	Total area	Built-up area in 1980	Built-up area in 1990	Built-up area in 2000	Built-up area in 2010
1980	13,002.92	9,744.78	10,148.41	10,796.63	11,316.19
1990	18,658.05	11,804.50	13,830.30	15,621.70	16,117.70
2000	28,813.03	13,355.06	16,366.43	21,083.79	21,849.19
2010	36,668.10	13,690.00	17,137.50	23,403.60	26,258.30

B. Bhatta, *Urban Growth Analysis and Remote Sensing*, SpringerBriefs in Geography,
DOI: 10.1007/978-94-007-4698-5,

Acronyms

CBD	Central business district
ETM+	Enhanced thematic mapper plus
GCP	Ground control point
GIS	Geographic information system
GPS	Global positioning system
IRS	Indian remote sensing
KMA	Kolkata metropolitan area
KMC	Kolkata municipal corporation
KMDA	Kolkata metropolitan development Authority
LISS	Linear imaging self-scanning sensor
MAD	Minimum average distance
MAUP	Modifiable areal unit problem
MSS	Multi-spectral scanner
SMSA	Standard metropolitan statistical Area
TM	Thematic mapper
UGB	Urban growth boundary

B. Bhatta, *Urban Growth Analysis and Remote Sensing*, SpringerBriefs in Geography,
DOI: 10.1007/978-94-007-4698-5,

Index

B. Bhatta, *Urban Growth Analysis and Remote Sensing*, SpringerBriefs in Geography,
DOI: 10.1007/978-94-007-4698-5, © The Author(s) 2012

MIX
Papier aus verantwortungsvollen Quellen
Paper from responsible sources
FSC® C105338

If you have any concerns about our products,
you can contact us on
ProductSafety@springernature.com

In case Publisher is established outside the EU,
the EU authorized representative is:
Springer Nature Customer Service Center GmbH
Europaplatz 3, 69115 Heidelberg, Germany

Printed by Libri Plureos GmbH
in Hamburg, Germany